# EV BATTERIES, CHARGERS, AND SUBSYSTEMS

For a complete listing of titles in the
*Artech House Power Engineering Library,*
turn to the back of this book.

# EV BATTERIES, CHARGERS, AND SUBSYSTEMS

Ashish Arora
Rita Garrido Menacho

ARTECH HOUSE
BOSTON | LONDON
artechhouse.com

**Library of Congress Cataloging-in-Publication Data**
A catalog record for this book is available from the U.S. Library of Congress.

**British Library Cataloguing in Publication Data**
A catalogue record for this book is available from the British Library.

**Cover design by Andy Meaden**

ISBN 13: 978-1-68569-051-9


**Artech House**
**685 Canton Street**
**Norwood, MA 02062**



10 9 8 7 6 5 4 3 2 1

*To my daughter Anya, my wife Dipali,*
*and my parents Neelima and Surender*
*—Ashish*

*To my parents Carmen and Gustavo,*
*my sister Andrea, and my fiancé Brady*
*—Rita*

# CONTENTS

# 8

# BATTERY-RELATED STANDARDS AND REGULATIONS 181



# 9

# VEHICLE-RELATED STANDARDS AND REGULATIONS 225

# PREFACE

As governments around the world have declared their intention to transition vehicle fleets on the road from gasoline-powered automobiles to automobiles running on alternative power sources over the next several decades, electric vehicles powered by lithium-ion (Li-ion) batteries have become the front-runner to replace gasoline-powered vehicles. Electric vehicles (EVs) powered by Li-ion batteries have become a common sight in almost all countries around the world. Li-ion batteries used in EVs, while much larger in size, are not fundamentally very different from the Li-ion batteries that power our cell phones, tablets, and other portable electronic devices. The architecture of all EVs on the road today has the same fundamental building blocks that consist of one or more electric motors connected to the wheels powered by a Li-ion battery. The need to charge the battery also requires EVs to be designed with on-board battery chargers and interface with external chargers. This necessitates the vehicles to be equipped with power conversion circuits that allow power to be transferred from one vehicle subsystem to another.

Chapter 1 of this book introduces the reader to the main EV subsystems with a high-level overview of what the subsystem does in the vehicle and how it interfaces with other EV subsystems.

Chapter 2 introduces the family of Li-ion cells. The discussion in Chapter 2 includes an overview of how Li-ion cells operate, the types of cells that are commonly found in EVs on the road today, and the steps involved in the manufacture of these cells. The chapter also focuses on aspects that are critical for EV applications: cell cycle life, end-of-life considerations, the need to address the potential of thermal runaway of Li-ion cells, and the types of abuse tests commonly performed on the cells to evaluate their safety and potential for thermal runaway. The chapter ends with a discussion of power sources that may be potential competitors to Li-ion cells in the future.

Li-ion batteries for EV applications typically incorporate hundreds of Li-ion cells surrounded by sophisticated electronic circuits that monitor and protect the cells and manage the battery thermal management circuits. The requirements and design architectures of Li-ion batteries in EV applications are detailed in Chapter 3 together with a review of functional safety requirements for the batteries. The chapter also provides an overview of algorithms used to provide drivers with an accurate state of charge of the batteries under different vehicle operating conditions. The chapter ends with a discussion of end-of-life recycling requirements for EV batteries and the concept of propagating circuit board failures and its importance in the context of EV batteries.

Chapter 4 provides an overview of Li-ion battery chargers while Chapter 5 discusses the types of motors that are potential candidates for use in EV applications. Both chapters provide an overview of the operation of the different charger and motor types and the advantages and disadvantages of each type together with an overview of some of the considerations that must be made when deciding the type of charger and motor to use in a specific EV application.

Power converters play a crucial role in EV applications as they allow the different EV subsystems to interface with each other and transfer power back and forth. While power converters have been around for a long time and are commonly found in a large number of applications, these converters have some unique requirements when used in EV applications. Chapter 6 introduces the reader to components used in power converters, the common power converter topologies suitable for use in EV applications, and the considerations that must be made when selecting a power converter topology for an EV application.

Chapter 7 introduces the reader to advanced driver assistance systems, their capabilities, and their limitations. While these systems are not unique to EVs, their ubiquity in EVs and growing adoption in conventional gasoline-powered vehicles makes them an important component of EVs on the road today. The chapter starts with an introduction of terminology associated with the levels of automation in vehicles and their benefits and then provides an overview of the sensors that are commonly used in these systems and the limitations of these sensors. The chapter ends with a discussion of the importance of functional safety when designing and developing these systems.

Chapters 8 and 9 guide the reader through standards and regulations that are applicable to EVs and their subsystems. The chapters also include a discussion of tests that are commonly performed on EVs and their subsystems to evaluate both reliability and safety. These chapters are not meant to be an exhaustive list of all the different tests and protocols followed by EV manufacturers around the world but include sufficient information to provide the readers with an overview of the tests and other considerations that must be made when developing or assessing EVs and their subsystems.

We are deeply indebted to the team at Artech for their constant support during the development of this book. Last but not least, we would like to acknowledge the unrelenting support from our families. Ashish would like to thank his daughter Anya, wife Dipali, and parents Surender and Neelima. Rita thanks her parents Carmen and Gustavo, sister Andrea, and fiancé Brady.

# 1

# INTRODUCTION

Global sales of electric vehicles (EVs) including plug-in hybrid EVs and fuel cell vehicles reached 13 million units in 2023, with a growth rate of 29.8% compared to 2022 [1]. While the growth rate has slowed down in 2024, this is considered a minor setback in the overall expected growth trajectory of EVs in the coming years. Most major automotive manufacturers offer one or more models of EVs for sale in the United States and in other countries around the world. While vehicle designs, performance specifications, driving range, and other features differ from one EV model to another, all EV applications have the same fundamental building blocks for operation. Most, if not all, EVs on the road today are equipped with the following, which will be discussed in detail in subsequent chapters of this book:

- A lithium-ion (Li-ion) battery that provides traction power;
- On-board and/or off-board chargers for replenishing the charge in the Li-ion battery;
- Motors connected to the wheels that use the energy from the battery to move the vehicle and provide energy to the battery, through regenerative braking, when the vehicle is slowing down;
- Power converters that are used for power conversion between the vehicle's traction battery and the motors;
- Advanced driver assistance systems (ADAS) that are designed to assist the driver under different vehicle operating conditions;

- Other subsystems that ensure adherence to safety regulations and other requirements that may vary from one country to another. These requirements include both mandatory crash-worthiness requirements and voluntary requirements such as safety certifications of EV components.

## 1.1 LI-ION BATTERIES

Benjamin Franklin is credited as being the first to use the term "battery" in relation to electricity which was the name he gave to his assembly of Leyden jars [2].[1] The term battery refers to a group of electrochemical cells that are devices that store energy in the form of potential energy and release it as electricity. The smallest electrochemical unit capable of providing power is called a cell. As the voltage of a single cell (nominally 3.6V to 3.8V for a Li-ion cell) is not sufficient to power an EV, individual cells are grouped to form modules and individual modules are grouped to form a battery in an EV application.

Li-ion batteries have become the exclusive choice of preference for battery chemistry for powering EVs due to their well-known advantages of comparatively higher-energy density, smaller size, and lower weight as compared to other competing battery chemistries. Li-ion battery energy capacity, measured in kilowatt-hours (kWh), can range from 40 kWh to over 100 kWh in EV applications making them from 4,000 to 10,000 greater in capacity as compared to a Li-ion battery in a typical cell phone today.[2]

Li-ion batteries for EV applications are designed with sophisticated battery management systems that perform a number of functions including determining the state of charge and state of health of the cells in the batteries. These sophisticated battery management systems are designed using the principles of functional safety[3] [3] to

1. With the Leyden jar, Benjamin Franklin found that "positive" charges resided on one side and "negative" charges resided on the other side of the jar. He discovered that, by linking multiple jars together, he could increase the amount of charge they could store. He called this new assembly of Leyden jars a battery.
2. A typical cell phone is equipped with a 10-Wh Li-ion battery.
3. At a high level, the term "functional safety" refers to the detection of a potentially dangerous condition resulting in the activation of a protective or

ensure that if a failure occurs, the batteries fail in a predictable manner to a "safe state."

## 1.2 CHARGERS

Charging EV batteries remains a pain point for EVs, as, unlike conventional gasoline vehicles, charging EV batteries can take a comparatively longer time. While charging times for EV batteries are steadily reducing, they remain considerably longer than the time it takes to fill the gas tank of a conventional gasoline vehicle. The charging time for EV batteries can range from 40 to 50 hours when using a Level 1 charger to less than 2 to 3 hours when using direct current (DC) fast chargers [4]. The lack of compatibility between EVs and some chargers adds to the complication, although this is being rectified with time [5].

EV battery chargers come in various designs depending upon how these chargers interface with the grid and the amount of power that the chargers can transfer at any one time. EV battery charging methods have also evolved with different charging technologies such as conductive and wireless charging. Other solutions being explored include battery swapping technologies that allow an EV user to replace their discharged battery with a fully charged battery at a designated station.

## 1.3 MOTORS

An EV requires a way of turning the electric energy from the vehicle's Li-ion battery into rotational motion required to move the EV forward. This is where the motor comes in. The role of a motor in an EV application is to convert electrical energy into mechanical energy to accelerate the vehicle and to convert mechanical energy into electrical energy when the vehicle is slowing down to capture the vehicle's energy and recharge the EV battery. There are many different motor types that can fulfill this requirement. At a high level, all electric motors regardless of their exact design use a stator that acts as the motor's stationary outer shell and a rotor that is the motor's rotating element.

corrective device or mechanism to prevent hazardous events from arising or providing mitigation to reduce the consequences of the hazardous event.

The stator's housing acts like an engine block of a conventional gasoline vehicle and is mounted to the vehicle's chassis while the rotor acts like the crankshaft of a conventional gasoline vehicle. Three different motor designs are suitable for use in an EV application. These include induction motors, permanent magnet motors, and brushless motors. The motor types can be broken down by key rotor differences that represent different ways of turning the stator's rotating magnetic field into actual rotational motion. The electric motor's power rating can range substantially from one motor to another. As an example, the motor in a Nissan Leaf EV is rated at 110 kW, while the Tesla Model S Performance motor is rated at 595 kW [6].

## 1.4 POWER CONVERTERS

The battery in an EV cannot be directly connected to the motor due to the different operating characteristics of the two subsystems. This is where power converters come in. Power converters in EVs allow the battery and the motor to work together. Following a similar working principle, EV chargers also use power converters to ensure the output power is appropriate for the battery. These converters can perform a number of tasks which include stepping up or stepping down the voltage, converting the voltage from DC to alternating current (AC) and vice versa, and changing the frequency of the AC voltage. Many different power converter designs and configurations are suitable for use in an EV application. Even though the power converter designs and architectures are different, the fundamental building blocks of these converters are common and include switching components such as metal oxide semiconductor field-effect transistors (MOSFETs), energy storage components such as inductors and capacitors, and current blocking components such as diodes. Because power converters in EV applications handle a large amount of energy during operation for years on end, they are designed with sophisticated cooling systems to prevent overheating and prolong life in the application.

## 1.5 ADVANCED DRIVER ASSISTANCE SYSTEMS

Just as the name suggests, advanced driver assistance systems (ADAS) are designed to assist the driver using sensors distributed around the vehicle to perceive the vehicle environment and either provide this information to the driver or take automatic action based on what is perceived. This action may include braking the vehicle, accelerating the vehicle, and steering the vehicle. The main goal of these systems is to reduce and/or prevent accidents by minimizing the amount of human interaction, and thus human error, needed through increased automation. While different vehicle models are equipped with different ADAS technology, some ADAS features will become mandatory in the coming years in some parts of the world. As an example, starting in 2029, vehicle manufacturers in the United States must make automatic emergency braking, which reduces vehicle and pedestrian crashes, standard in cars and light trucks. This technology will require all cars to stop and avoid contact with a vehicle in front of them up to 62 mph and for cars to detect pedestrians in both daylight and darkness. The standard will also require the systems to brake automatically up to 90 mph when a collision with a lead vehicle is imminent, and up to 45 mph when a pedestrian is detected [7].

## 1.6 STANDARDS AND REGULATIONS

Federal Motor Vehicle Safety Standards (FMVSS) are US regulations that specify minimum safety requirements to be met by vehicles and vehicle components and include crash avoidance, crashworthiness and damage susceptibility, and post-crash survivability requirements. EVs are also required to meet additional electrolyte spillage and electric shock protection related requirements specifically under crash scenarios. In addition, while not mandatory, a number of standards have been developed around the world for both EV batteries and chargers. These standards drafted by organizations such as the Society of Automotive Engineers (SAE), the International Electrotechnical Commission (IEC), and Underwriters Laboratories (UL) are commonly used to evaluate the safety of EV batteries and chargers. EV battery safety standards define a series of mechanical, electrical, thermal/environmental, and cell internal short-circuit related tests to evaluate the response of the EV batteries to abuse conditions expected in the field. Similarly, standards focused on EV chargers define both requirements to ensure protection against thermal incidents and electrical shock hazards.

## References

[1] Lu, M., "Visualizing Global Electric Vehicle Sales in 2023, by Market Share," *Visual Capitalist*, March 9, 2024, https://www.visualcapitalist.com/visualizing-global-electric-vehicle-sales-in-2023-by-market-share.

[2] Smithsonian Institution, "Benjamin Franklin's Pioneering Electrical Work Influenced Today's Technology," https://www.si.edu/newsdesk/releases/benjamin-franklins-pioneering-electrical-work-influenced-todays-technology.

[3] Mendias, M., A. Arora, and S. Lele, "Functional Safety & Safety Critical Systems - An Overview," SAE Technical Paper 2021-01-0157, 2021.

[4] US Department of Transportation, "Charger Types and Speeds," https://www.transportation.gov/rural/ev/toolkit/ev-basics/charging-speeds.

[5] Barry, K., "Automakers Move to a Common Plug Standard to Allow Their EVs to Use Tesla Superchargers," *Consumer Reports*, May 1, 2024, https://www.consumerreports.org/cars/hybrids-evs/tesla-superchargers-open-to-other-evs-what-to-know-a9262067544/.

[6] Bonnici, D., "EV 'kW' Figures Explained," *WhichCar*, January 22, 2022, https://www.whichcar.com.au/car-advice/what-does-ev-kw-mean.

[7] NHTSA, "NHTSA Finalizes Key Safety Rule to Reduce Crashes and Save Lives," April 29, 2024, https://www.nhtsa.gov/press-releases/nhtsa-fmvss-127-automatic-emergency-braking-reduce-crashes.

# 2

# LITHIUM-ION CELLS

Most EVs offered for sale today use Li-ion as the battery technology for traction power. This is also true for most if not all hybrid EVs sold today. Li-ion cells, first commercialized in the 1990s, refer to a family of battery chemistries that can serve as hosts for lithium ions. Billions of Li-ion cells power a range of devices today from toys used by toddlers to commercial aircraft flying people around the world to EVs. The adoption of Li-ion cells in applications around the world has occurred due to the many well-known advantages of the cells over other cell chemistries. Table 2.1 demonstrates the reason that Li-ion cells are the natural choice of battery chemistry in all applications including EVs [1]. As compared to the other competing battery chemistries (e.g., lead acid and nickel metal hydride (NiMH)), Li-ion cells have a superior energy density and thus can store a greater amount of energy in a smaller package, making them ideally suited as a power source in all applications where battery size and weight are important considerations.

Other fundamental advantages of Li-ion cells include the reduction potential of lithium. Lithium has the lowest reduction potential of any element (–3.04V), allowing lithium-based cells to have one of the highest cell potentials.[1] One advantage of Li-ion cells over competing cell chemistries is its relatively higher nominal voltage.

1. In theory, the highest potential for a single elemental cell would be achieved using lithium (standard electrode (reduction) potentials in aqueous solution at 25°C = –3.04V) and fluorine (standard electrode (reduction) potential in aqueous solution at 25°C = +2.87V), resulting in a single cell with an open-circuit

**Table 2.1**
Comparison of Different Battery Technologies

| Property | Unit of Measurement | Lead Acid | NiMH | Li-Ion |
|---|---|---|---|---|
| Cell voltage | Volts | 2 | 1.2 | 3.2–3.7 |
| Energy density | Wh/Kg | 30–40 | 50–80 | 100–500 |
| Power density | W/Kg | 100–200 | 100–500 | 500–8,000 |
| Useful capacity | Depth of discharge % | 50 | 50–80 | 100 |
| Cycle life | Number of cycles | 600–900 | >1,000 | >2,000 |

*From:* [1].

Typical Li-ion cells used have a nominal voltage rating that ranges from 3.2V to 3.7V depending on the exact Li-ion cell chemistry. However, regardless of the Li-ion cell chemistry, the nominal voltage of a Li-ion cell is comparatively much higher than the 1.2V nominal voltage rating of a NiMH cell. In addition, lithium is the third lightest element and has one of the smallest ionic radii of any single charged ion allowing lithium-based cells to have a comparatively higher gravimetric and volumetric capacity and power density [2]. These many advantages make Li-ion chemistry a natural choice for powering EVs.

## 2.1 LI-ION CELL OPERATION

At a high level, a Li-ion cell is a cell in which the negative electrode (anode) and the positive electrode (cathode) materials serve as a host for lithium ions as shown in Figure 2.1 [3] (the terms anode and cathode will be used to refer to the negative electrode and the positive electrode, respectively, in this chapter and in the remainder of the book).

In practice, a Li-ion cell is composed of various layers of anode and cathode materials that can be stacked or rolled creating alternating layers (a detailed description of the cell manufacturing process is provided later in this chapter). These alternating layers of anode and cathode materials are separated by a porous film (separator). An electrolyte provides the media for the transport of lithium ions between the layers. The most common varieties of Li-ion cells use graphite-based and lithium metal oxide-based active materials that are coated

voltage of 5.91V. However, due to the extremely active nature of the two elements, as a practical matter, such a cell is currently not available.

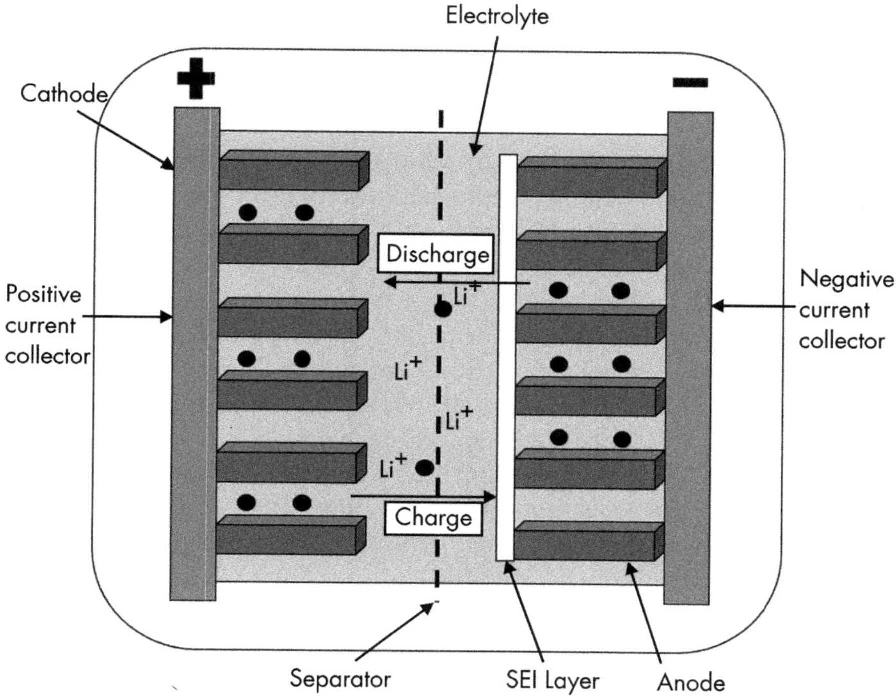


**Figure 2.1** Li-ion cell operation.

on the surface of the negative and positive current collectors, respectively. The electrolyte is commonly an organic solvent containing a dissolved lithium salt.

The operation of a Li-ion cell involves the movement of lithium ions from the positive to the negative electrode when the cell is being charged. The process is reversed as a cell is discharged. The lithium ions are reversibly inserted and removed from the electrodes through a process called intercalation. Intercalation does not result in any significant changes to the electrode. To increase the Li-ion cell's power capability, it is desirable for the anode and cathode materials to have large geometric electrode areas with high porosity to increase the reaction area. For this reason, Li-ion cell electrodes are constructed of a paste composed of fine particles coated on thin current collectors (usually thin copper foil for anode or aluminum foil for cathode) [4].

The choice of electrode materials within the cell and the selection of the electrode/electrolyte combination has been a topic of intensive research and development with Li-ion cells specifically for their use in EVs. This choice (often referred to as the Li-ion cell chemistry), differs based on the requirements of the application and has a significant impact on capacity, life and safety. Table 2.2 [5] lists some of the different materials used for the positive electrode (cathode), the negative electrode (anode), and the electrolyte in Li-ion cells. Some of the materials (e.g., NMC, LFP, and NCA) are more suitable for use in EVs as compared to some other available Li-ion cell chemistries (e.g., LCO and LMO).

The primary function of a separator in a Li-ion cell is to prevent physical contact between the anode and the cathode while facilitating Li-ion transport within the cell. Separators in Li-ion cells are also required to be tolerant to abuse conditions, stable at voltages above 4V, and chemically inert to other cell materials. Most commercially available nonaqueous separators for Li-ion cells are single-layer or multilayer polymer sheets typically made of polyolefins such as polyethylene (PE) or polypropylene (PP), which have transition temperatures typically between 135°C and 165°C. Multilayer separators for Li-ion cells are also designed with a shutdown feature where two of the layers have different phase transition temperatures. This results in the lower melting temperature component filling the pores of the other solid layer and stopping the ion transport and hence the current flow in the cell as the temperature of the cell increases [6]. Li-ion cells manufactured for EV applications also often use ceramic-coated separators where a layer of ceramic coats one side or both sides of the separator with the aim of improving the stability and abuse tolerance of the cells.

## 2.2 CELL TYPES

Li-ion cells come in a variety of form factors. Prismatic, cylindrical, and pouch cells are the common configurations of Li-ion cells used in EVs.

### 2.2.1 Prismatic Cells

Prismatic cells are designed with a rectangular metallic enclosure that is typically made of aluminum or steel. The metallic enclosure is sealed via a welding process and serves as a heat dissipater. Prismatic

**Table 2.2**
Some Types of Li-Ion Cells

| Type | Lithium Cobalt Oxide ($LiCoO_2$) | Lithium Manganese Oxide ($LiMn_2O_4$) | Lithium Nickel Manganese Cobalt Oxide ($LiNi_xMn_yCo_{1-x-y}O_2$) | Lithium Iron Phosphate ($LiFePO_4$) | Lithium Nickel Cobalt Aluminum Oxide ($LiNi_xCo_yAl_zO2$) | Lithium Titanate ($Li_4Ti_5O_{12}$) |
|---|---|---|---|---|---|---|
| Abbreviation | LCO | LMO | NMC | LFP | NCA | LTO |
| Anode | Graphite carbon | Graphite carbon | Graphite carbon | Graphite carbon | Graphite carbon/ part silicon | Lithium titanate |
| Cathode | Lithium cobalt oxide | Lithium manganese oxide | Lithium nickel-manganese-cobalt oxide | Lithium iron phosphate | Lithium nickel cobalt oxide | Lithium manganese oxide |
| Notes | Relatively high capacity but limited specific power<br>Cobalt is expensive<br>Serves as an energy cell<br>Market share has stabilized | High power but less capacity<br>Safer than Li-cobalt (LCO)<br>Commonly mixed with NMC to improve performance | Provides high capacity and high power<br>Energy cell/power cell hybrid<br>Favorite chemistry for many applications | Very flat voltage discharge curve but low capacity<br>Safety version<br>Comparatively higher self-discharge<br>Used primarily for energy storage, forklifts, bicycles | Shares similarities with Li-cobalt (LCO)<br>Serves as an energy cell | Long life, fast charge, wide temperature range but relatively low capacity and expensive<br>Among the safest Li-ion cells<br>Ability to ultra-fast-charge<br>High-cost limits adoption |

*From:* [5].

cells come in different sizes and are typically designed with a vent that is often (but not always) positioned on the top of the cell between the cell terminals. Figure 2.2 shows an example of a commercially available prismatic cell for EVs. Prismatic cells have a higher-energy density due to their compact form factor and are typically available in sizes up to 150 Ah.

### 2.2.2 Cylindrical Cells

One of the first commercially available Li-ion cells were manufactured as cylindrical cells. Cylindrical cells are referred to by their dimensions. As an example, a 21700 cylindrical cell refers to a cell that has a diameter of 21 mm and a length of 70 mm. Similarly, an 18650 cylindrical cell refers to a cell with a height of 65 mm and a diameter of 18 mm. Over time, many different dimensions of cylindrical cells have been manufactured for different applications.

The metal can of a cylindrical cell forms the cell's negative terminal with the positive terminal and vent at the top of the cell (also referred to as the cell's cap). Because the cell can is always energized, cylindrical cells are often sold partially covered with shrink-wrap

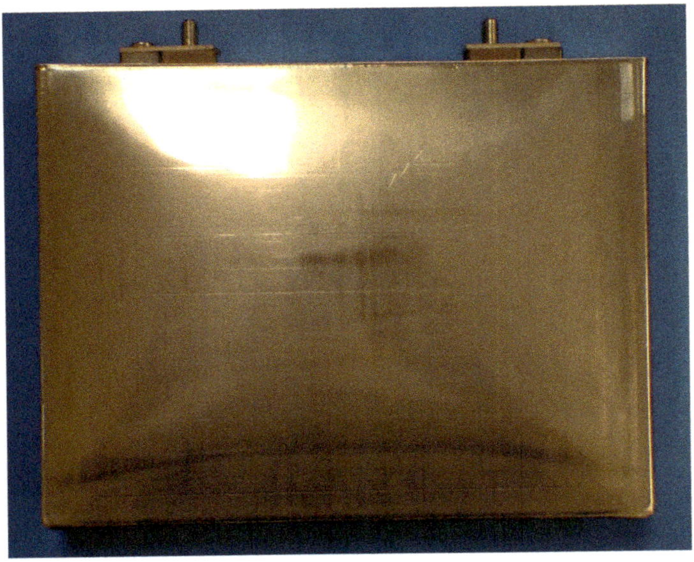

**Figure 2.2** Commercially available prismatic cell.

that provides electrical isolation. Cylindrical cells use a single "jelly roll" of electrodes and separator and often utilize a current interrupt device (CID) for safety.[2] The can seal is typically formed through a crimp [7, 8].

Figure 2.3 shows the profile view of an 18650 cylindrical cell along with the cell's cap. Cylindrical Li-ion cells come in sizes up to 20 Ah. The cap assembly of a cylindrical cell contains a number of protection elements. These include a burst disk that acts as the venting mechanism for the cell. Due to the form factor of the cells, battery packs made with cylindrical cells cannot achieve the same packaging density as battery packs made with prismatic cells [9]. Figure 2.4 shows the electrode assembly (often called the jelly roll) of a commercially available 18650 cylindrical cell.

### 2.2.3 Pouch Cells

Pouch cells have become very popular specifically in portable consumer electronic applications due to their flat configuration and flexible packaging. Pouch cells have a relatively higher energy density and are produced in different sizes based on their intended application. The relatively higher packaging density makes them specifically desirable for use in EV applications. One common construction of the pouch cell is the packaging of a jelly roll in a pouch with metallic tabs attached to the positive and negative electrodes protruding from the pouch. Figure 2.5 shows a commercially available pouch cell with

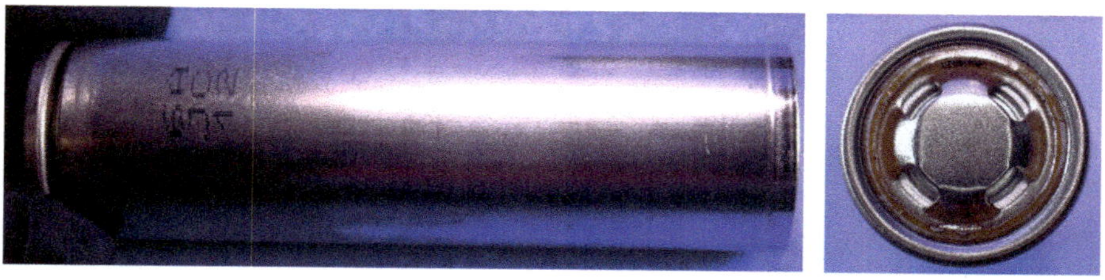

**Figure 2.3** Profile view of a commercially available cylindrical cell (left) and the cell's cap (right).

2. Unlike cylindrical cells that are typically used in lower-voltage portable consumer electronic devices, cylindrical cells in EV applications are not typically designed with a positive temperature coefficient (PTC) device. This is because the higher nominal voltage of EVs (several hundred volts) and the larger required discharge currents makes the PTC unsuitable for application.

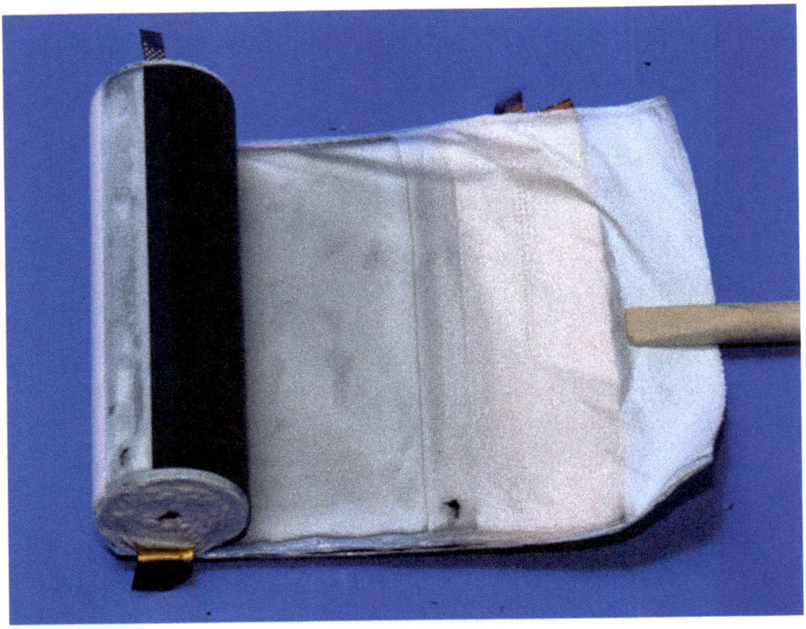

**Figure 2.4** Electrode assembly of a commercially available 18650 cylindrical cell.

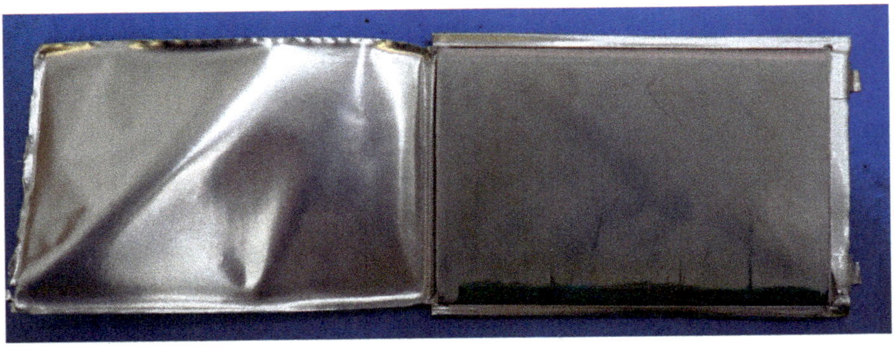

**Figure 2.5** Commercially available pouch cell.

the enclosure cut off to reveal the electrodes. Pouch cells also include electrodes and a separator that are folded, stacked, or wound. Pouch cells generally vent at the sealed edge of the pouch [8].

## 2.3 LI-ION CELL MANUFACTURING PROCESS

The manufacturing process for a typical Li-ion cell [10] involves several controlled steps during three distinct phases of manufacturing. These phases are: (1) electrode manufacturing, (2) cell assembly, and (3) cell finishing.

### 2.3.1 Electrode Manufacturing

The electrode manufacturing process is where the anode and cathode for a Li-ion cell are manufactured. Figure 2.6 shows the typical steps involved in the electrode manufacturing process.

#### *2.3.1.1 Mixing*

The assembly line for the manufacturing of the cathode and anode is almost always separated at a cell manufacturing facility. This is done to prevent cross-contamination between the anode and cathode, which not only can affect the performance of the manufactured cells but can also have safety implications. Electrode manufacturing (of both the anode and the cathode) starts with a mixing process, which involves the preparation and blending of various components into a homogeneous mixture that will form the electrodes.

Anodes for Li-ion cells almost always use graphite as the active material. The mixing process involves adding additives (typically conductive carbon), solvents (e.g., deionized water), and a binder to the active material. Carboxymethyl cellulose (CMC) is often used as a binder for the active material. Other additives (e.g., styrene-butadiene (SBR)) are also sometimes added to the active material. The first step of the mixing process involves a dry mixing of the various materials in a large drum. Once the mixing is complete, additional solvents are added to allow for a dispersion of the wet active material into a tank for the next step in the assembly process. The wet active material pumped to the tank is in the form of a slurry or paste. Variations in the electrode mixing process can occur depending on the battery manufacturer, type of battery, or even the scale of production.

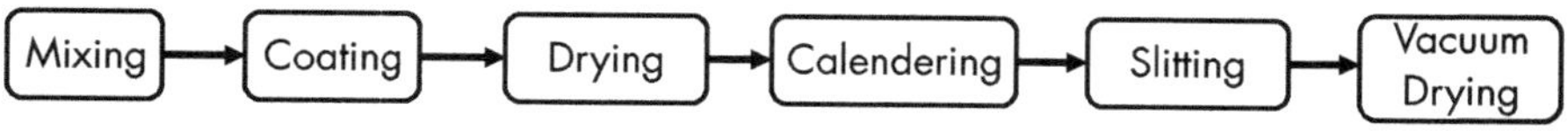


**Figure 2.6** Electrode manufacturing process.

The cathode mixing process is almost identical and uses one of the many options of a layered transitional metal oxide depending upon the type of Li-ion cell being manufactured. Similar to the anode mixing process, additives (typically carbon black), solvents (e.g., N-Methyl-2-Pyrrolidone or NMP), and a binder material (e.g., polyvinylidene fluoride or PVDF) are added to the active material.

#### *2.3.1.2 Coating*

The next step in the electrode manufacturing process called coating involves a process where the anode and cathode slurry is deposited on foils. A copper foil is used for the anode and an aluminum foil is used for the cathode. Large rolls of aluminum or copper are unrolled on a conveyer system and an applicator tool is used to apply the slurry to the unrolled foils. The coating of the active material can be continuous or intermittent depending upon the type of cell being manufactured. The top and bottom sides of the foils are often coated sequentially although newer Li-ion cell manufacturing lines have simultaneous coating of both sides of the foil. The coating thickness depends upon the cell design but is often in the 50 to 100-$\mu$m range. The electrode coating process requires precise control of various parameters such as coating thickness, coating speed, and drying conditions to ensure consistent electrode layers. Quality control checks are typically performed during and after the coating process to verify the coating thickness, uniformity, and overall electrode quality.

#### *2.3.1.3 Drying*

The coated foils go through a drying system that is implemented using a series of connected chambers to remove the solvent from the electrode mixture and leave behind a solid and adherent electrode layer on the foil (or current collector). The drying system often involves a series of ovens each set to a different temperature to achieve the required consistency for the coated active material. After drying, the foil is cooled down and then can be coated on the second side if needed or rewound for the next step in the assembly process. Figure 2.7 shows a typical drying process for the coated foils.

#### *2.3.1.4 Calendaring*

The drying process is followed by calendaring, which involves a pair of heavy rollers that compress the coated aluminum/copper foils (Figure

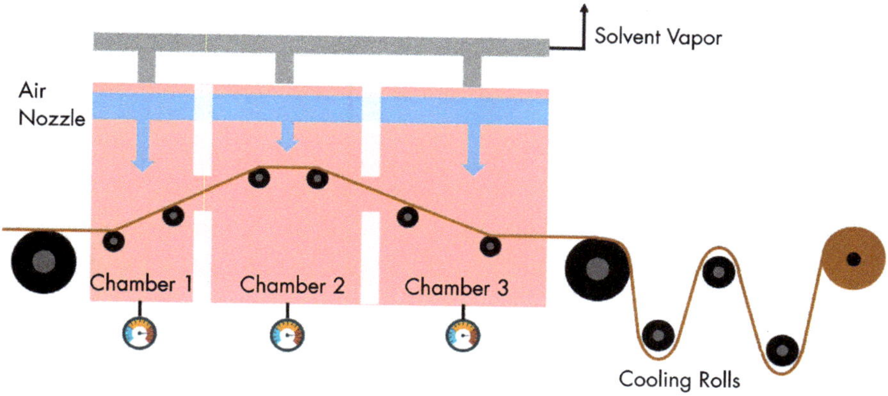


**Figure 2.7** The drying process. (*After:* [10].)

2.8) and remove any excess solvents. The purpose of calendaring is to help with the alignment of the particles within the electrode materials and improve the overall density and consistency of the electrode, which is necessary to ensure the performance and stability of the cell. The desired thickness and density specifications for the electrodes can be controlled by adjusting the rollers of the calendaring machine. Brushes or an air/flow system is often used to clean the electrodes before and after the compression. The electrode foil is rolled up again after calendaring for the next step of the assembly process. Vision systems are often used to measure the thickness of the compressed active material to ensure consistency.

#### 2.3.1.5 Slitting

Calendaring is followed by slitting, which prepares the electrode for further assembly into battery cells. Slitting involves cutting and shap-

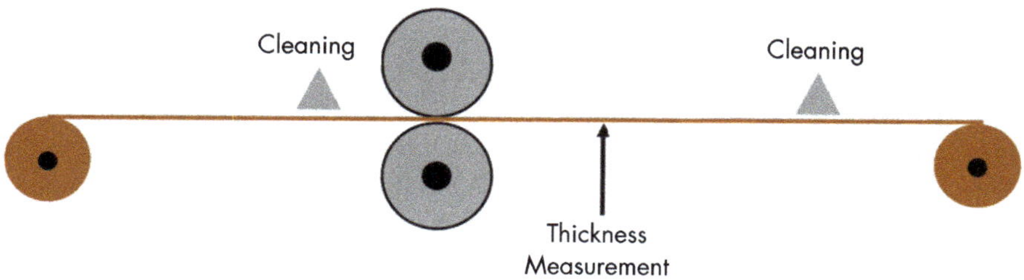


**Figure 2.8** Calendaring.

ing the coated and dried electrodes to the desired size and configuration for the specific cell design.

#### *2.3.1.6 Vacuum Drying*

The last step in the electrode manufacturing process is vacuum drying where the coated and slit electrode rolls are stored in vacuum ovens to remove any residual moisture and solvents, which can impact the performance of the cells in the field. The lowering of the air pressure allows the effective evaporation of any moisture and other volatile substances at a lower temperature. The vacuum drying time can range from 12 to 30 hours depending upon the type of electrode being assembled. The electrodes are transferred to a dry room or dry-packed under vacuum once the vacuum drying process is complete.

### 2.3.2 Cell Assembly

The cell assembly process starts with combining the electrodes together. This can be done through a winding or a stacking process. These two processes differ in how the electrode layers, the separator and the electrolyte, are arranged within the cell. Both the winding and stacking processes have their own advantages and disadvantages and the suitability of a process depends upon factors such as cell size and energy density requirements.

#### *2.3.2.1 Cell Winding*

In a cell winding process, the electrode layers and the separator are rolled or wound together to form a cylindrical or oval-like structure. This winding is typically done using automated equipment that applies controlled tension to ensure uniform and secure winding. Figure 2.4 shows an example of a wound cylindrical cell.

#### *2.3.2.2 Cell Stacking*

In a cell stacking process, the electrode layers and separator are stacked in a flat configuration to form a prismatic or pouch-shaped cell. Figure 2.9 shows a computed tomographic (CT) scan of a Li-ion pouch cell with a stacked electrode assembly. The electrode stacking process requires proper control to ensure that the electrodes are not damaged during the process, the anode covers the cathode completely

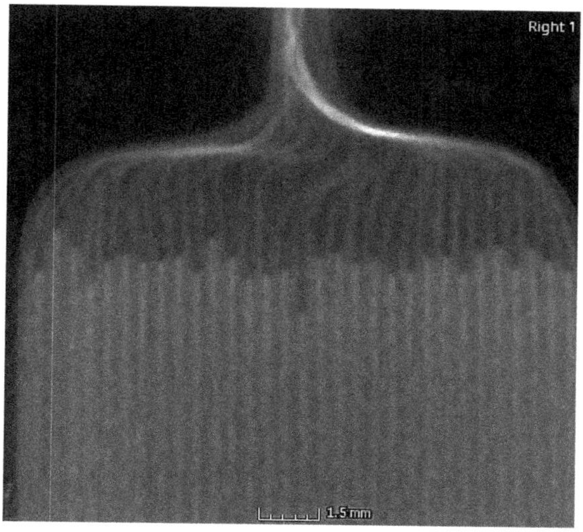


**Figure 2.9** Electrode stacking.

and no damage or other shifting of the separator occurs during the process.

Another electrode stacking process known as the Z-folding process involves the insertion of individual electrode sheets into a folded separator film. The anode and cathode electrodes are individually placed onto a continuous separator film, which is folded over the inserted electrode allowing the next electrode to be positioned into the electrode assembly. This process is repeated until the electrode assembly is completed. Some of the claimed advantages of the Z-folding process over the more conventional electrode stacking process include the lower probability of the electrodes making contact at the outer edges, better alignment of the electrodes during assembly, and faster assembly of the electrodes into cells.

*2.3.2.3 Packaging*

The wound or stacked electrode assembly is then packaged into an enclosure. This enclosure can be a metal container for a cylindrical or prismatic cell or a pouch for a pouch type cell. As an example, for a cylindrical cell, the wound electrode assembly or jelly roll is inserted into a metal container and the current collector for the anode is typically welded to the bottom of the container while the current collector for the cathode is welded to the lid to provide electrical connection between the container and the electrodes. An insulation ring is used on both the top and bottom of the jelly roll to insulate the jelly roll from the container. A similar process is used when assembling prismatic cells.

#### 2.3.2.4 *Electrolyte Filling*

The packaging process is followed by electrolyte filling where electrolyte is added to the electrode assembly that was inserted into its enclosure. A high-precision dosing needle is used to add electrolyte to the cell under vacuum. A pressure profile is applied to the cell during the filling process through a supply of inert gas and/or generation of a vacuum in alternating operation to activate the capillary effect in the cell (called wetting). The evacuation and partial filling may be repeated several times depending upon the type of cell being assembled. The electrolyte-filling process is followed by sealing the cell.

### 2.3.3 Cell Finishing

The cell finishing process involves three steps: formation, aging, and end-of-line checking.

#### 2.3.3.1 *Cell Formation*

Cell formation (also called initial charging) is an important process during the manufacturing of a Li-ion cell and involves the initial charging and conditioning of the cell to activate its electrochemical properties and optimize its performance. The formation process is required to wet the liquid electrolyte into the porous electrodes and can take several days at elevated temperatures. During the formation process, the cells are placed in racks and are often contacted by spring-loaded contact pins to charge and discharge the cells using well-defined voltages and currents. This allows the lithium ions to embed into the porous structure of the graphite in the anode and create a necessary protective layer called the solid electrolyte interface (SEI) layer between the electrolyte and the negative electrode. The SEI layer plays an important role in limiting the self-discharge of the cell and improving the performance and life of the cell [11].

Additionally, the first charge cycle of larger pouch cells can generate a significant amount of gas in the cell. This gas is pressed out of the cell into a dead space called the gas bag. Degassing of this gas bag is done by piercing the pouch in a vacuum chamber and sucking the gases away from the cell. The cell is then finally sealed under vacuum after the gas bag is separated from the cell.

#### *2.3.3.2 Cell Aging and End-of-Line Checking*

The cell formation step is followed by cell aging, which is one of the final steps in the cell manufacturing process and is used for quality assurance. Aging involves subjecting the cells to accelerated aging tests to simulate long-term usage and assess the performance of the cells. Aging is followed by end-of-line checking or final testing where key performance parameters of the cells (e.g., capacity, voltage, internal resistance) are measured. Cells that meet the required specifications and quality standards move on to subsequent manufacturing steps, while those that do not meet the criteria may be rejected or undergo further analysis and corrective actions.

## 2.4 CELL CYCLE LIFE AND END-OF-LIFE CONSIDERATION

Unlike in portable consumer electronic applications where a typical Li-ion cell is expected to power the device for 3 to 4 years, Li-ion cells in EVs are required to power the vehicle for 10 or more years. This requirement makes it necessary to understand the cycle life of the cell that is chosen for the application and when the cell is expected to reach its end of life. All batteries age and lose capacity as they do. For an EV, this means a shorter driving range with time.

Li-ion batteries age with use (cycle life) and they age with time (calendar life), even if not used. Although these two aging mechanisms at a high level can be thought of as separate, in an EV application, just as in all other applications, they act as overlapping forces. Depending upon the type of application and usage pattern, one may become more dominant than the other. Environmental conditions such as temperature and humidity also influence the two aging mechanisms in their own ways. Finally, the actual chemistry of the Li-ion cell also determines how the cells age in terms of cycle and calendar life.

### 2.4.1 Cycle Life

The capacity a Li-ion cell reduces every time the cell is charged and discharged. Many applications define the end of life of a Li-ion cell as the point at which the capacity of the cell drops to less than 80% of its nominal capacity.[3] At this point, the cell is still capable of storing a significant amount of its rated capacity, but the rate of capacity degradation starts to increase and the overall performance of the cell reduces. Before the cell reaches its end of life, the capacity degradation of a cell typically decreases linearly with each charge and discharge cycle. A number of factors can impact the capacity degradation of a Li-ion cell. These include the depth of discharge, the charge and discharge current magnitudes, the operating temperature, and so on. As an example, Figure 2.10 [3] shows the capacity degradation as a function of charge/discharge cycles for a Li-ion cell with a lithium cobalt dioxide positive electrode and a graphite negative electrode. The

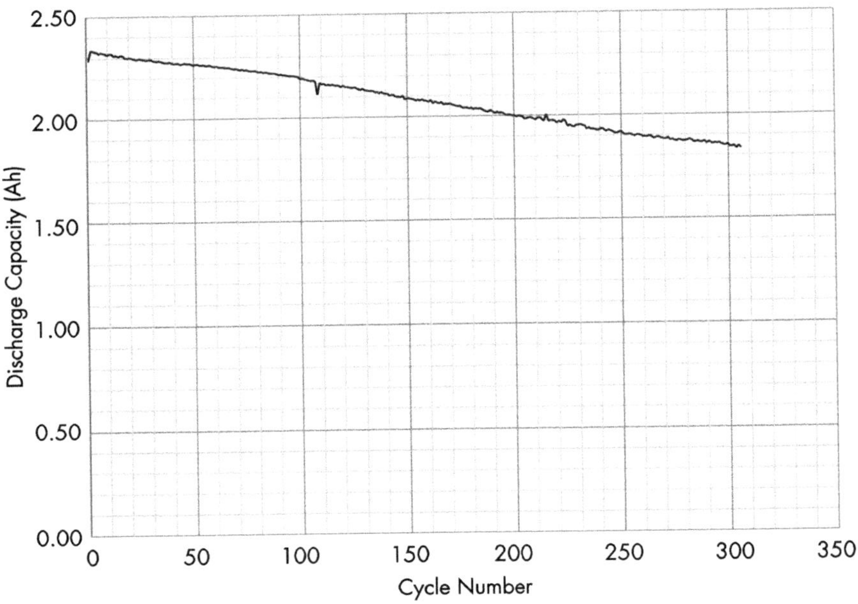


**Figure 2.10** Capacity as a function of charge/discharge cycles of a commercially available Li-ion cell.

3. Some cell manufacturers define end of life as the point at which the capacity drops to 70% of its nominal capacity.

figure indicates that the cell loses approximately 20% of its rated capacity after 300 charge/discharge cycles under the tested conditions at an ambient temperature of 25°C.

### 2.4.2 Calendar Aging

Calendar aging refers to the degradation of a Li-ion cell's capacity when the cells are not being charged or discharged and just sitting on a shelf. The ambient temperature at which a cell operates and/or is stored can have a large impact on the cell's capacity degradation. Factors that affect calendar aging include:

- *Temperature:* Elevated temperatures generally accelerate the calendar aging process in a Li-ion cell. Cell datasheets typically suggest that cells be stored at an ambient temperature below 20°C, especially for longer periods of time (several months). A typical rule of thumb is that every 10°C increase in the ambient storage temperature of a Li-ion cell doubles the rate of capacity degradation due to calendar aging.
- *State of charge (SOC):* Capacity degradation due to calendar aging is also dependent on the cell SOC during storage. Higher SOCs increase calendar aging especially if the SOC during storage is above 60%. For this reason, Li-ion cells are often stored at partial states of charge (typically around 40% to 60%) for long-term storage.

Figure 2.11 [12] shows the normalized capacity degradation of a fully charged cylindrical Li-ion cell with a lithium cobalt oxide cathode and a graphite anode at an ambient temperature of 60°C. As can be seen in Figure 2.11, calendar aging is a function of cell SOC during storage.[4]

### 2.4.3 End of Life

As is obvious in the terminology itself, the end of life of a Li-ion cell in the context of an EV application refers to the state reached by the

4. While Figure 2.11 may suggest that storing the cell at 0% SOC may be ideal to minimize calendar aging, this is never done in practice. This is because Li-ion cells self-discharge during storage. Storing Li-ion cells at 0% SOC for prolonged periods of time increases the risk of cell over-discharge which can damage the cell.

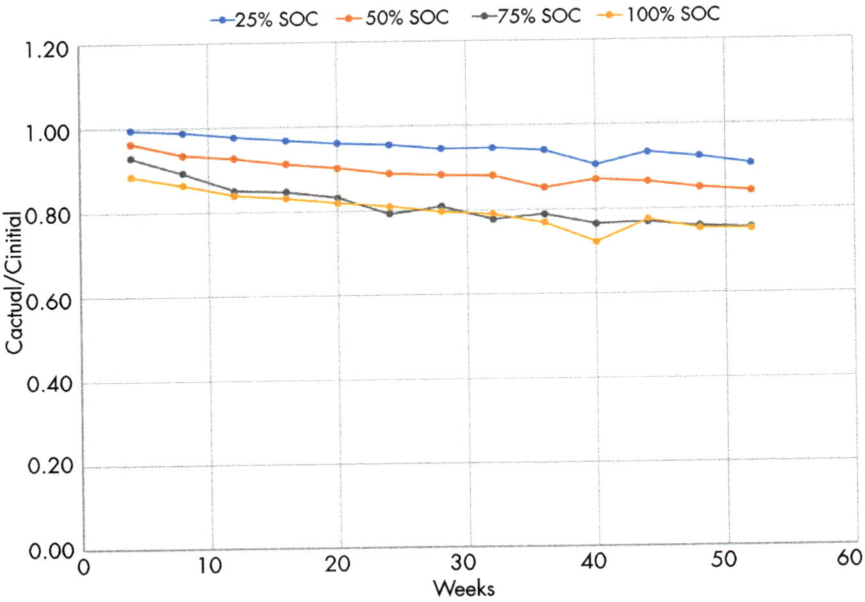


**Figure 2.11** Calendar aging of a polymer Li-ion cell with a lithium cobalt oxide cathode and a graphite anode.

Li-ion cell where it can no longer serve its function and is defined using some measure of change in performance from the cell's performance when it was initially manufactured. End-of-life criteria differ for Li-ion cells used in energy and power applications. When determining the capacity of a Li-ion cell or battery for an application, it is important to consider both power and energy ratings for a cell/battery and to select a cell that will meet the application's energy and power requirements for the expected life of the product. For Li-ion cells used in energy applications, the cell's end of life is commonly taken as the point when the cell's usable energy drops to 70% to 80% of its rated capacity. Determining whether a Li-ion cell or battery is suitable for the application requires identifying the end-of-life criteria for the cell/battery and predicting the capacity degradation of the cell/battery due to both cycle and calendar aging. The actual aging of a Li-ion cell in an EV application will depend upon a number of factors such as charge level, depth of discharge, ambient temperature during operation, and storage. As an example, Table 2.3 [13] shows the ex-

**Table 2.3**
Discharge Cycles and Capacity as a Function of Charge Voltage for an NMC Cell

| Cell Charge Level | Discharge Cycles | Available Stored Energy (%) |
|---|---|---|
| 4.20 | 300–500 | 100 |
| 4.15 | 400–700 | 90–95 |
| 4.10 | 600–1,000 | 85–90 |
| 4.05 | 850–1,500 | 80–85 |
| 4.00 | 1,200–2,000 | 70–75 |
| 3.90 | 2,400–4,000 | 60–65 |

*From:* [13].

pected discharge cycles and capacity as a function of charge voltage for an NMC cell.

## 2.5 THERMAL RUNAWAY IN LI-ION CELLS

A failure of a Li-ion cell in an application due to any reason is almost always nonenergetic (benign). This may be exhibited as a cell that can no longer be charged or discharged, a cell that loses capacity in an accelerated manner or a cell that is permanently disabled due to the operation of its internal protection component (e.g., the current interrupt device in a cylindrical cell). In rare instances, a Li-ion cell may experience a more energetic failure and fail exothermically. This is often referred to as the cell going into thermal runaway. The thermal runaway of a Li-ion cell is a condition where the heat generated within the cell exceeds the heat dissipated by the cell. The thermal runaway of a Li-ion cell can occur due to a number of reasons and may result in the cell surface temperature exceeding 800°C. Some of the factors that can cause a Li-ion cell's failure are as follows:

- *External short-circuit:* Exposing a Li-ion cell to an external short-circuit condition can result in high cell discharge currents, specifically for cells with relatively lower internal impedances. The high discharge current causes relatively rapid heating of the cell, which, under some conditions, can cause the cell to vent. The rapid increase in a cell's internal pressure can also result in a rupture of the cell enclosure.

- *Cell manufacturing defects:* Cell manufacturing defects increase the likelihood of a development of a cell internal short-circuit condition, which can result in a cell failure in the field. For this reason, Li-ion cells should be assembled using a manufacturing process that is controlled and precise. Examples of cell manufacturing defects introduced during the manufacturing process include improper cell tab positioning, winding misalignments, bending of the positive or negative electrodes during the manufacturing process, introduction of metallic/nonmetallic contaminants into the cell during the assembly process, improper electrode coating, and damage to the electrodes during assembly.
- *Cell charging algorithm:* Li-ion cells have a well-defined specification for charging voltage and current. The cells are charged using a constant-current-constant-voltage (CC-CV) charge profile. In this charge profile, a constant current is used to charge the cell until the cell voltage reaches 4.2V.[5] Once the cell voltage reaches 4.2V, the voltage is maintained and the charge current starts to drop. Charging is terminated once the charge current drops below a certain threshold (typically 0.01C). Li-ion cells require a precise charge protocol to be followed to ensure that the cells are not charged outside their rated specifications.
- *Cell overcharge:* Once a Li-ion cell reaches its fully charged condition, any additional charge current will result in an overcharging of the cell. This current starts to heat the cell and, if continued, increases the likelihood of the cell becoming thermally unstable and going into thermal runaway.
- *Cell over-discharge:* Li-ion cells cannot be discharged to 0V during operation. Discharging the cell below its specified discharge cutoff voltage (2.3V to 2.7V for LCO cells) can result in the oxidation of the current collector in the negative electrode. This oxidation leads to copper dissolution in the electrolyte. The dissolved copper redeposits in regions of the cell and is reduced back to copper metal when the over-discharged cell is charged again. Frequent over-discharging of the cell can result in the formation of copper dendrites between the cell elec-

5. Some cell chemistries can be charged to a voltage as high as 4.4V.

trodes, which increases the likelihood of a cell internal short-circuit.

- *Operation outside rated temperature:* Li-ion cells have a limited temperature range of operation. For LCO cells this temperature is from 0°C to 45°C for charging and from –20°C to 60°C for discharging. Operating the Li-ion cell outside its specified temperature range increases the probability of a cell failure in the field.

## 2.6 OTHER POWER SOURCES

### 2.6.1 Fuel Cells

Fuel cells are another source of often discussed traction power for EVs. Fuel cell vehicles are yet to gain traction in the market with the Toyota Mirai being one of the few fuel cell passenger vehicles available for purchase today in the United States. Unlike Li-ion or other rechargeable batteries, fuel cells do not need to be charged. Fuel cells can generate electricity as long as supplied by a fuel (typically hydrogen). The fuel cell anode is fed by the fuel (hydrogen) while air (oxygen) is fed to the cathode to enable the fuel cell reaction that produces electricity [14].

Just like Li-ion cells, fuel cells also come in many types. However, the basic operation of all fuel cells is the same. Different electrolytes define the various types of fuel cells. Just like the Li-ion cell types described in Table 2.2, the various types of fuel cells have their advantages, which determine their application. However, the basic operation of all fuel cells is the same and involves an oxidation at the anode and a reduction at the cathode with the two reactions connected by a charged species that migrates through the electrolyte while the electrons flow through the external circuit. As an example, in a polymer electrolyte membrane fuel cell, a catalyst separates hydrogen atoms into protons and electrons, which take different paths to the cathode. The electrons go through an external circuit, creating a flow of electricity. The protons migrate through the electrolyte to the cathode, where they reunite with oxygen and the electrons to produce water and heat [15]. Existing challenges with fuel cells have limited their desirability for EVs.

The adoption of fuel cells for vehicular applications has been hampered by the limited fueling infrastructure (hydrogen fuel stations). Given the limited production of fuel cells, the industry has yet to enjoy the benefits of economies of scale, which has resulted in still relatively high costs for certain materials used in fuel cells. One other disadvantage of fuel cells when compared to Li-ion batteries is the comparatively lower nominal voltage for the cells which in practical applications is less than 1.0V.

### 2.6.2 Solid-State Batteries

At a high level, the operating principle of solid-state batteries is not very different from Li-ion cells. Just like in Li-ion cells, solid state batteries contain two electrodes, and their operation involves the movement of ions through the electrolyte between the electrodes. The main difference between Li-ion cells and solid-state batteries involves the electrolyte. Unlike in Li-ion cells, solid-state batteries use a solid electrolyte. This eliminates the need for a separator between the cathode and anode, which is required in Li-ion cells due to the liquid electrolyte. Some of the claimed advantages of solid-state batteries over Li-ion cells include the ability to withstand comparatively higher and lower temperatures, the ability to charge much faster than Li-ion cells, a comparatively longer life span and a greater flexibility in form factor for the cells [16]. However, solid-state batteries still need to overcome several challenges to act as a viable alternative to Li-ion cells for EV applications. These challenges include developing and manufacturing solid electrolytes that can maintain proper contact with the electrodes in the cells, developing electrodes that can store comparatively more energy than Li-ion cell electrodes and developing manufacturing processes that allow the batteries to be manufactured at the scale required for EV applications [16].

## 2.7 LI-ION CELL ABUSE TESTING

Li-ion technology is constantly evolving toward maximizing energy density while optimizing functionality and safety. As described in previous sections, though thermal runaway is a rare failure mode of Li-ion cells, it is the most damaging and harmful to property and individuals. As such, safety considerations and hazard mitigations are implemented into Li-ion cells at various stages of their life cycle. During manufacturing, ceramic coatings can be applied to the separator to serve as a reinforcer and allow the separator to withstand higher mechanical damage and temperatures. Additionally, cylindrical Li-ion cells manufactured for EV applications contain a current-interrupt device (CID). A CID is an internal protective device serving as a permanent fuse in the event of a high temperature, pressure, or voltage. When cells are embedded in EVs or other consumer products, sophisticated battery management systems (BMS) accompany them to serve as monitoring systems. A state-of-the-art BMS monitors the state of charge and state of health of the cells and mitigates against overcharge, over-discharge, over-temperature, under-temperature, and cell imbalance conditions by disabling the system (the BMS will be discussed in detail in Chapter 3).

Furthermore, Li-ion cells and battery packs undergo rigorous testing before they become available to consumers. Standards and regulations are frequently updated to test and validate Li-ion cells and battery packs under abusive conditions that they might encounter in the field. The aim is to characterize the cells' response to these conditions, understand associated risks, pinpoint vulnerabilities, and identify areas of improvement with the overall goal of minimizing the occurrence of thermal runaway. As a result, various test protocols state as their passing criteria that no fire and/or explosion shall be observed during the test and/or after the test for a predetermined amount of time (monitoring period) ranging from a few hours to days. A detailed description of safety standards and regulations is provided in Chapter 8.

The tests listed in various standards can be broadly broken up into three categories. These categories are electrical, mechanical, and environmental [17, 18]. The external short-circuit test is an example of an electrical test meant to evaluate the effectiveness of the BMS and/or cells in handling the large current generated due to an external short-circuit condition. When performed at the pack level, the test is meant to characterize the activation of the overcurrent protection in the BMS to assess if appropriately set for the cells. If performed at the stack/cell level, the test evaluates if the cell or cells can withstand the current and heat generated without reaching hazardous conditions such as venting, electrolyte leakage, explosion, or flames. The test is performed by applying a short-circuit of a certain resistance for a specified amount of time to the cell or pack terminals. The magnitude of the output resistance required to simulate a short-circuit condition varies among standards but typically falls in the range of a few to tens of m$\Omega$. Figure 2.12 shows an example test setup for a short-circuit test performed on a cylindrical 18650 cell. The instrumentation for this test typically includes measuring the short-circuit current, the cell voltage and the cell temperature during the test.

Another example of a mechanical abuse test commonly performed for cells used in EV applications is the nail penetration test. The goal of this test is to simulate a cell internal short circuit condition. During the test, a metallic nail of a specific material, diameter, and tip angle is embedded into the cell at a specific speed. Once the nail penetrates through the various anode and cathode layers of the cell, a conductive path is created simulating an internal short circuit between the layers. Finegan et al. [19] described the variation in cell behavior when undergoing nail penetration and provided insights into breakdown mechanisms during this process and how these can lead to the initiation and propagation of thermal runaway. Figure 2.13 shows an example test fixture for a nail penetration test. Most nail penetration test fixtures contain a hydraulic or pneumatic-based system that allows for precise control of the nail penetration speed and the depth of penetration during the test. The instrumentation for this test typically includes measuring the cell voltage and the cell temperature during the test.

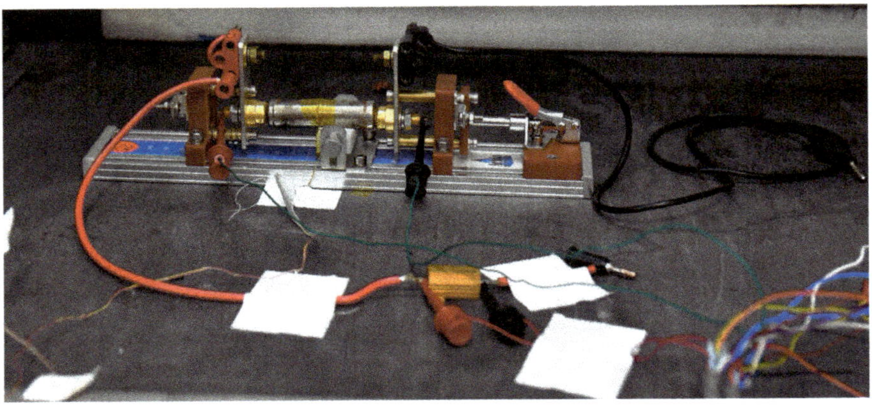

**Figure 2.12** Example test set-up for a short-circuit test.

**Figure 2.13** Example test set-up for a nail penetration test.

Lastly, the altitude test, which is part of the UN 38.3 standard, is an example of an environmental abuse test that simulates transport in an unpressurized airplane cargo area at a 15,000-m altitude. The test consists of placing the cells in a vacuum chamber set for 11.6 kPa for approximately 6 hours. The passing criteria for this test include no electrolyte leakage, mass loss, venting, rupture, fire, or a voltage drop of more than 10% of the pretest voltage. Altitude testing is performed to determine if the cells can handle an unplanned loss in pressure in the cargo area during air transport without displaying any hazardous conditions.

As described, thermal runaway exposes individuals and property to elevated temperatures and fire. In addition to these conditions, failing cells could also vent and/or leak electrolyte exposing individuals to corrosive, toxic, and flammable fumes and fluids. The gases released during venting of a cell may include a mixture of carbon monoxide, carbon dioxide, methane, hydrogen, oxygen, ethane, ethylene, and hydrofluoric acid [20–24]. Due to the hazardous nature of these fumes, various standards have developed testing and evaluation procedures to quantify the concentration of gases released during abuse testing. The gas levels measured should be below the concentrations defined by organizations such as the Occupational Safety and Health Administration (OSHA) and the Environmental Protection Agency (EPA).

## References

[1] Johnson Matthey Battery Systems, "Our Guide to Batteries," 2012, http://www.jmbatterysystems.com/JMBS/media/JMBS/Documents/JMBS-Guide-to-Batteries.pdf.

[2] Nitta, N., et al., "Li-Ion Battery Materials: Present and Future," *Materials Today*, Vol. 18, No. 5, June 2015.

[3] Arora, A., et al., *Lithium-Ion Battery Failures in Consumer Electronics*, Norwood, MA: Artech House, 2019.

[4] Mikolajczak, C., et al., "Lithium-Ion Batteries Hazard and Use Assessment," Final Report, *NFPA,* July 2011.

[5] van Schalkwijk, W., and B. Scrosati, *Advances in Lithium-Ion Batteries*, New York: Springer Science & Business Media, 2007.

[6] Orendorff, C. J., "The Role of Separators in Lithium-Ion Cell Safety," *The Electrochemical Society Interface*, Summer 2012.

[7] Reddy, T. B., (ed.), *Linden's Handbook of Batteries*, 4th ed., New York: McGraw-Hill, 2011.

[8] Weicker, P., *A System's Approach to Lithium-Ion Battery Management*, Norwood, MA: Artech House, 2014.

[9] Maiser, E., "Battery Packaging - Technology Review," *AIP Conference Proceedings*, Vol. 1597, 2014, p. 204.

[10] Heimes, H., et al., "Lithium-Ion Battery Cell Production Process," Third Edition, Rwthaachen University/VDMA, 2018.

[11] Davoodabadi, A., et al., "Effect of Calendaring and Temperature on Electrolyte Wetting in Lithium-Ion Battery Electrodes," 2019, https://www.osti.gov/servlets/purl/1761757.

[12] Lele, S., A. Arora, and K. Benson, "Predicting the Life of Li-Ion Batteries Using the Arrhenius Model," *Battcon*, Nashville, TN, 2018.

[13] BU-808: How to Prolong Lithium-Based Batteries - Battery University, https://batteryuniversity.com/article/bu-808-how-to-prolong-lithium-based-batteries.

[14] https://www.energy.gov/eere/fuelcells/fuel-cell-basics.

[15] https://www.montanagreenpower.com/other/fuel_cells.html.

[16] https://article.murata.com/en-us/article/basic-lithium-ion-battery-4.

[17] Lai, X., et al., "A Review of Lithium-Ion Battery Failure Hazards: Test Standards, Accident Analysis, and Safety Suggestions," *Batteries*, Vol. 8, No. 11, November 20, 2022.

[18] Ruiza, V., et al., "A Review of International Abuse Testing Standards and Regulations for Lithium Ion Batteries in Electric and Hybrid Electric

Vehicles," *Renewable and Sustainable Energy Reviews*, Vol. 81, Part 1, January 2018, pp. 1427–1452.

[19] Finegan, D. P., et al., "Tracking Internal Temperature and Structural Dynamics During Nail Penetration of Lithium-Ion Cells," *Journal of The Electrochemical Society*, Vol. 164, No. 13, October 31, 2017.

[20] Golubkov, A. W., et al., "Thermal Runaway Experiments on Consumer Li-Ion Batteries with Metal-Oxide and Olivin-Type Cathodes," *RSC Adv.*, Vol. 4, 2014, pp. 3633–3642.

[21] Campion, C. L., W. Li, and B. L. Lucht, "Thermal Decomposition of LiPF 6-Based Electrolytes for Lithium-Ion Batteries," *Journal of The Electrochemical Society*, Vol. 152, 2005, pp. A2327–A2334.

[22] Lecocq, A., et al., "Comparison of the Fire Consequences of an Electric Vehicle and an Internal Combustion Engine Vehicle," *Proceedings of the International Conference on Fires in Vehicles – FIVE*, Chicago, IL, 2012, pp. 183–194.

[23] Roth, E. P., and D. H. Doughty, "Thermal Abuse Performance of High-Power 18650 Li-Ion Cells," *Journal of Power Sources*, Vol. 128, No. 2, April 5, 2004, pp. 308–318.

[24] Ohsaki, T., et al., "Overcharge Reaction of Lithium-Ion Batteries," *Journal of Power Sources*, Vol. 146, 2005, pp. 97–100.

# 3

# EV BATTERIES

## 3.1 REQUIREMENTS AND DESIGN ARCHITECTURES

Li-ion cells have a well-defined range of operating voltages, currents, and temperatures that are typically specified in the cell datasheet by the manufacturer. Designing a battery requires ensuring that the cells in the battery do not operate outside these well-defined ranges either under normal operating conditions or in the event of a fault. This is where the battery management system (BMS) comes in. The BMS typically incorporated within an EV battery performs a number of important functions. These include ensuring that the cells operate within their rated specifications, disabling the battery pack in the event of a failure, accurately estimating the SOC of the battery pack and providing this information to the driver. Most, if not all, EV battery designs utilize multiple modules connected to each other. Each module consists of a number of cells connected in series and parallel along with a monitoring board. The monitoring board typically performs the following functions:

- Monitors the voltage of every cell in the module;
- Monitors the temperature at selected locations within the module;
- Balances the series connected cell groups;
- Communicates with the central BMS in the battery pack.

Figure 3.1 shows the typical architecture of an EV battery. Each of the four modules incorporate a monitoring board, which communicates with the BMS[1] that performs the following functions:

- Communicates with the module monitoring boards;
- Disables battery pack charging and discharging if information from any of the module monitoring boards indicates that a cell or monitored temperature is outside the rated specifications (e.g., overcharge and over-discharge);
- Monitors the charge and discharge current to prevent battery operation outside the rated specifications;
- Protects the battery from external abuse conditions (e.g., short-circuit)
- Determines the battery SOC;
- Communicates with the vehicle and provides information, such as the state of health and the SOC of the battery.[2]

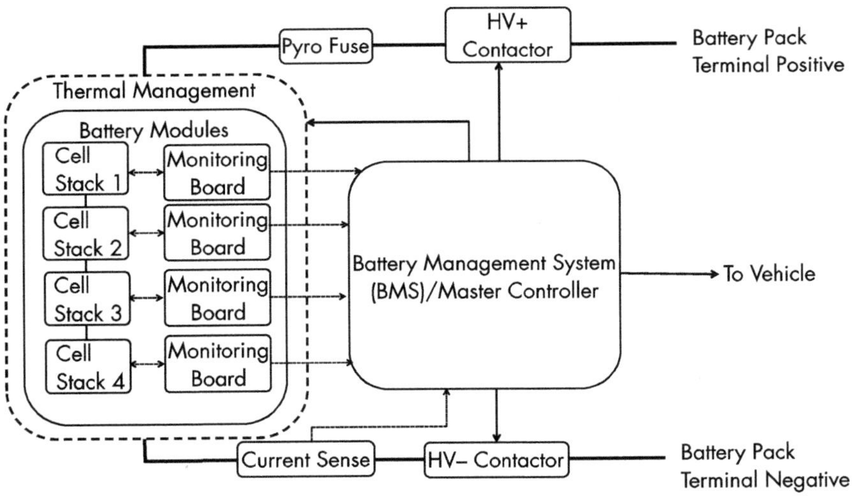


**Figure 3.1** Typical EV battery architecture.

1. Central BMS and BMS will be used interchangeably throughout the chapter when in the context of EV applications.
2. EV batteries are also designed to disable their output in the event of a crash. In these designs, the vehicle communicates the occurrence of a crash to the central BMS. The central BMS opens the contactors to disable the battery pack on receiving information about a crash from the vehicle.

The flow of charge and discharge current into and out of the battery is controlled through multiple contactors in the battery. The central BMS can open the contactors to disable the battery and prevent it from charging or discharging.[3]

Figure 3.2 shows an example of a traction battery that is equipped with 294 pouch-type Li-ion cells, while Figure 3.3 shows an example of a traction battery equipped with cylindrical Li-ion cells.

### 3.1.1 Functional Safety and EV Batteries

An EV battery is an example of a safety-critical system and needs to be designed as such [2]. A safety-critical system is a system in which human safety is dependent upon the designed operation of the system. Most safety-critical systems, including EV batteries, are designed to

**Figure 3.2** Example of a traction battery equipped with pouch-type Li-ion cells.

3. Some EV battery pack designs also incorporate a pyrofuse in series with the contactors for over-current protection.

**Figure 3.3** Example of a traction battery equipped with cylindrical Li-ion cells.

achieve functional safety. The term “functional safety” addresses the concept of implementing protective or corrective mechanisms that are activated upon the detection of a potentially dangerous condition in an application. The protective/corrective mechanisms mitigate/reduce the consequences of the potential hazardous events. This results in a system design that ensures that the system transitions to a safe state of operation on the detection of a failure to reduce the consequences of a potentially hazardous event. The designed safe state of operation ensures the safety of users, maintenance personnel, or other stakeholders of the application.

The concept of functional safety involves the implementation of active systems rather than passive systems. These active systems are designed to trigger a predesigned fail-safe function upon detecting a hazardous condition. The detection of smoke by sensors in a building that results in the activation of fire sprinklers is an example of an active safety system. The activation of the “limp home” mode of operation in an automobile due to the detection of a failure in one of the accelerator pedal position sensors to prevent vehicle malfunction is another example of an active safety system. The concept of functional safety excludes any safety achieved through the use of passive safety systems such as fire-protection blankets in an EV around the vehicle’s traction battery. In addition, design features in vehicles such as rounded (i.e., not sharp) corners on door handles and other vehicle locations contacted by drivers on a regular basis are not examples of hazards mitigated through functional safety principles.

A number of different industries have developed functional safety standards to provide guidelines on the identification of, and system response to, faults/failures. As an example, IEC 60601 and IEC 62304 are often used as functional safety standards for medical devices, and IEC 62061 is used in the machine tooling industry. At a high level, functional safety standards require that a risk posed by the equipment under control (EUC) and its control system should be identified, analyzed, and tested against tolerability criteria and all risks found to be intolerable must be reduced. The risk reduction method may involve changing the actual design of the EUC. If a design change is not feasible, safety functions must be incorporated either within the control system or in one or more protection systems for risk reduction. The functional safety standard developed by the automotive industry and commonly used for the design and development of EV batteries is ISO 26262 (Road Vehicles – Functional Safety) [1].[4] ISO 26262 is a standard for functional safety of electrical and/or electronic systems in production automobiles that was first released in 2011. ISO 26262 defines functional safety as the "absence of unreasonable risk due to hazards caused by malfunctioning behavior of a system" [1].

The goal of all functional safety standards including ISO 26262 is risk reduction. Risk reduction can be achieved by means of safety functions that are based on an understanding of the risks. In the context of EV batteries, the goal of these safety functions is to achieve the reduction of the risk that is posed by the battery system to a risk that is deemed in the given circumstances to be acceptable.[5] The idea behind the various functional safety standards is the development of safety functions for risk reduction. These safety functions are developed through the following steps:

- Identification and analysis of risks;
- Evaluation of the tolerability of the identified risks;
- Determination of the required risk reduction for each identified risk;

4. IEC 61508 was the first developed functional safety standard and has been used as the starting point for several subsequently developed functional safety standards including ISO 26262.
5. Acceptable or tolerable risk is the amount of risk that is considered to be acceptable for a given context and typically varies from one application to another.

- Specification of the safety requirement for each risk reduction, including their safety integrity levels (SILs);
- Design and implementation of safety functions to meet the safety requirements;
- Validation of the safety functions.

#### *3.1.1.1 Risk Identification, Analysis, and Assessment*

One of the first steps in achieving functional safety of an EV battery is a risk analysis, which involves the identification of the risk posed by the battery system. The risk analysis process consists of the following three stages:

- Hazard identification that involves the identification of a potential source of harm activated either internally or externally that can result in material or life-threatening conditions. For EV batteries, hazards typically fall into the following six categories:
  - Electrical hazards: short-circuit, overcharge;
  - Thermal hazards: fire, elevated temperatures;
  - Mechanical hazards: shock, vibration, drop, penetration;
  - Environmental hazards: floods, earthquakes;
  - System hazards: transients imposed on utility power supplies, failures of inverters/chargers;
  - Chemical hazards: leaking cells.
- Hazard analysis that involves understanding the chain of cause and effect between the identified hazards and their consequences. The analysis should determine the causes and consequences such that the risk associated with each hazard, which may be quantitative or qualitative, can be identified. If a quantitative approach is taken for risk analysis, the probabilities of events are estimated and assigned numeric values for their consequences. The risk is then calculated by multiplying the two.
- Risk assessment involves comparing the risk values determined during the hazard analysis against tolerability criteria[6]

6. The tolerability criteria are typically determined at the start of the development process by the team performing the hazard analysis.

to determine if the identified risks are tolerable and, if not, by how much they need to be reduced.

Functional safety standards provide flexibility in the use of tools for the hazard identification and analysis process. Some of the commonly used tools during the hazard analysis process include:

- Failure modes and effects analysis (FMEA), which is designed to identify potential failure modes for a product or process and assess the risk associated with the identified failure modes. Once the risks are identified, a ranking system is used to flag the most serious concerns that require the implementation of corrective actions. FMEA teams typically identify the threshold criteria for the ranked risks and require corrective actions for any identified risks that exceed the preset threshold criteria.
- Hazard and operability study (HAZOP), which involves a structured technique for system examination and risk management. This is often used for identifying potential hazards in a system and identifying operability problems likely to lead to nonconforming products.
- Process hazard analysis (PHA), which involves a systematic approach for identifying, evaluating, and controlling the hazards of processes. The methodology establishes priority based on:
  - Extent of process hazards;
  - Number of potentially affected individuals;
  - Age of process;
  - Operating history of process.
- Fault tree analysis (FTA), which provides a means to logically and graphically display the paths to failure for a system or component.

#### 3.1.1.2 *Safety Requirements, Functions, and Safety Integrity Levels*

Once the risk analysis is performed, the safety requirements for the battery can be defined for the purpose of risk reduction. At the design stage, the safety requirements are provided by means of safety functions that are implemented in safety-related systems. As an example,

Table 3.1 [2, 4] lists some safety functions that may be identified for EV batteries in response to the safety requirement of preventing a fire. Similar safety functions may be developed for other potential battery-related safety requirements, such as electrolyte leakage and venting of toxic gases.

The safety integrity level (SIL) is a concept that measures the rate of unsafe failures. IEC 61508 (Functional Safety of Electrical/Electronic/Programmable Electronic Safety-Related Systems) is one of the first developed functional safety standards and defines SIL as "the likelihood of a safety-related system satisfactorily performing the required safety functions under all the stated conditions, within a stated period of time" [3]. The standard further defines a SIL as "a discrete level (one of four) for specifying the safety integrity requirements of safety functions" [3]. In other words, a SIL is a target probability of failure of a defined safety function. As an example, Table 3.2 lists the average probability of failure per hour for continuous operations at the various SILs per IEC 61508.

As previously mentioned, the functional safety standard used for automotive systems including EV battery packs is ISO 26262 (Road Vehicles – Functional Safety). In ISO 26262, the classification of hazards is done through a systematic process, which combines factors of severity, probability, and controllability into overall Automotive Safety Integrity Levels (ASILs). Each of these variables is broken down into subclasses. Severity has four classes ranging from "no injuries" (S0) to "life-threatening/fatal injuries" (S3). Exposure has five classes covering the "incredibly unlikely" (E0) to the "highly probable" (E4).

**Table 3.1**
Examples of Safety Functions for EV Battery Packs Used to Prevent a Fire

| Safety Function | Description |
|---|---|
| 1 | Cell overcharge protection |
| 2 | Battery overcharge protection |
| 3 | Battery short-circuit and over-current protection |
| 4 | Battery reverse polarity protection |
| 5 | Protection against charging of over-discharged battery |
| 6 | Protection against charging of over-discharged cell |
| 7 | Protection against over temperature on battery |

*From:* [2, 4].

**Table 3.2**
Safety Integrity Levels for Continuous Operation as Defined in IEC 61508

| SIL | Average Probability of Failure per Hour |
|---|---|
| 4 | $\geq 10^{-9}$ to $< 10^{-8}$ |
| 3 | $\geq 10^{-8}$ to $< 10^{-7}$ |
| 2 | $\geq 10^{-7}$ to $< 10^{-6}$ |
| 1 | $\geq 10^{-6}$ to $< 10^{-5}$ |

Controllability has four classes ranging from "controllable in general" (C0) to "uncontrollable" (C3). All variables and subclassifications are analyzed and combined to determine the required ASIL (as shown in Figure 3.4). ASILs can range from the low-risk ASIL A to the highest-risk ASIL D. As an example, a combination of the highest hazards (S3 + E4 + C3) would result in an ASIL D classification. An additional Quality Management (QM) designation, falling below ASIL A, is also available, which indicates that the manufacturer may address the lowest-risk hazards according to an alternative process in lieu of conforming to the functional safety standard.

The functional component and the safety integrity component are defined upon identification of the safety requirement. As discussed previously, the goal of the functional component is to reduce the risk.

| | | Probability | Controllability | | |
|---|---|---|---|---|---|
| | | | C1 | C2 | C3 |
| Severity | S1 | E1 | QM | QM | QM |
| | | E2 | QM | QM | QM |
| | | E3 | QM | QM | A |
| | | E4 | QM | A | B |
| | S2 | E1 | QM | QM | QM |
| | | E2 | QM | QM | A |
| | | E3 | QM | A | B |
| | | E4 | A | B | C |
| | S3 | E1 | QM | QM | A |
| | | E2 | QM | A | B |
| | | E3 | A | B | C |
| | | E4 | B | C | D |

**Figure 3.4** ASIL determination.

The safety integrity component consists of an ASIL that is related to the amount of risk reduction that is required. The greater the risk reduction needed, the greater the extent to which safety depends on the system that provides the risk reduction, so the higher the SIL for the safety requirement. Example ASILs for the safety functions defined in Table 3.1 are shown in Table 3.3 [2]. The ASILs shown in Table 3.3 make certain assumptions about the consequences of the identified risk. As an example, the table assumes that the consequence of a battery-level overcharge/short-circuit will be comparatively more severe than a cell-level overcharge/short-circuit. This is generally true because in an EV battery, a battery-level abuse condition can result in damage to multiple cells simultaneously. For this reason, the safety functions associated with battery-level protection specifically in EV applications are provided higher ASIL ratings as compared to safety functions associated with protection at the individual cell level [4].

The definition of the safety functions is followed by the identification of safety requirement specifications that are provided to the battery designers so that they can be implemented in the battery design. Depending upon whether the safety requirements are to be

**Table 3.3**
ASIL Assignments for Safety Functions Identified to Achieve a Safety Goal of Preventing a Fire in an EV Battery Pack

| Safety Function | Risk | Protection Threshold | ASIL |
|---|---|---|---|
| 1 | Cell overcharge | 4.25V ± 0.1V (primary)<br>4.35V ± 0.3V (secondary) | A |
| 2 | Battery overcharge | 420V ± 5.0V (primary)<br>435V ± 2.0V (secondary) | C |
| 3 | External short-circuit | 500A ± 5A for 10 ms (primary)<br>Fuse 750A (secondary) | C |
| 4 | Reverse polarity | 500 Vdc (primary) | A |
| 5 | Charge after over-discharge of battery | 270V ± 1.0V (primary)<br>250V ± 1.0V (secondary) | B |
| 6 | Charge after over-discharge of cell | 1.9V ± 0.1V (primary)<br>1.5V ± 0.1V (secondary) | A |
| 7 | Battery over-temperature | 60°C ± 3°C (primary)<br>75ºC ± 3ºC (secondary) | B |

*From:* [2].

implemented using hardware components or software algorithms, specific hardware and software requirements that form part of the overall battery design are then defined. Once the design is completed, extensive testing is required to validate that the implemented design meets the defined safety requirements. The level of testing required is directly dependent upon the ASIL of the safety function and is often performed by independent safety assessors.

A similar approach can also be utilized when applying software fail-safe techniques to ensure adequate fail-safe coverage. ISO 26262 discusses the application of certain hazard analysis and software fail-safe techniques applied to an electric brake system. These techniques identify hazards such as "failure to provide desired acceleration consistent with driver intent" and its causes such as "controller failure." To investigate the effectiveness of strategies used to detect possible controller failures, a coverage matrix is developed, and each potential controller failure is assigned a severity level. A test suite is then applied to the design to mitigate the risk of each element causing a potential hazard.

#### *3.1.1.3 The Safety Case*

The eventual goal of the functional safety standard is for the development of a safety case, which is a safety argument for a system backed by a sufficient body of evidence demonstrating that the system is adequately safe to be used in a particular application. Traceability is shown from the design requirements and specifications to the development, verification, and validation of safety functions. Although the safety case is not currently mandatory, many organizations use this as a work product for the functional safety assessment process.

## 3.2 BATTERY THERMAL MANAGEMENT SYSTEMS

The high charge and discharge currents coupled with a wide range of ambient temperatures can result in a relatively wide range of operating temperatures for the cells in an EV battery pack. A lack of control of the cell operating temperatures can accelerate cell capacity degradation, which can have an adverse impact on an EV's driving range. For this reason, EV battery packs require a thermal management system to ensure that the cells operate within their rated temperature and that there is a relatively uniform distribution of temperature in the battery pack. Most EV battery packs incorporate sophisticated battery thermal management systems designed with both cooling and heating mechanisms to achieve this goal. EV battery thermal management systems can use a combination of techniques for thermal management. These typically include [5–7]:

- Air cooling;
- Liquid cooling;
- Phase change material cooling.

As the name suggests, air cooling systems use air as the medium for the cooling infrastructure. Air-cooled systems are low-cost, simple, and low weight, which results in low manufacturing costs. However, air-cooled systems can have limited effectiveness due to the relatively low air thermal conductivity. Specifically when operating at higher ambient temperatures, air-cooled systems struggle to regulate the temperatures and consequently are not as effective at maintaining a uniform temperature across the battery pack [6]. Air-cooled systems in EV battery packs can utilize passive or active air cooling [7].

In a passive air-cooled EV battery pack, air from the surroundings of the vehicle and/or cabin is used to heat or cool the battery pack with the relative motion of the vehicle aiding the air flow. The use of this "ambient" air for thermal management can have limited effectiveness as it is dependent on the environmental conditions surrounding the vehicle. EV battery packs utilizing active air cooling still use air from the vehicle surroundings or cabin (Figure 3.5). However, these systems employ the heating, ventilation, and air conditioning (HVAC) capabilities within the vehicle to cool or heat the battery pack [8–10].

Liquid-cooled EV battery packs use coolants such as water and/or ethylene glycol for temperature management of EV battery pack cells. Most EV battery packs using liquid cooling are designed with metallic coolant lines/plates that run the full length of the battery pack ensuring that every cell in the battery pack is in physical contact with the coolant carrying infrastructure. The BMS monitors and controls the coolant intake and output temperatures, coolant flow rate, and battery pack temperatures for optimal operation. Liquid cooling is a common temperature management technique used in EV battery

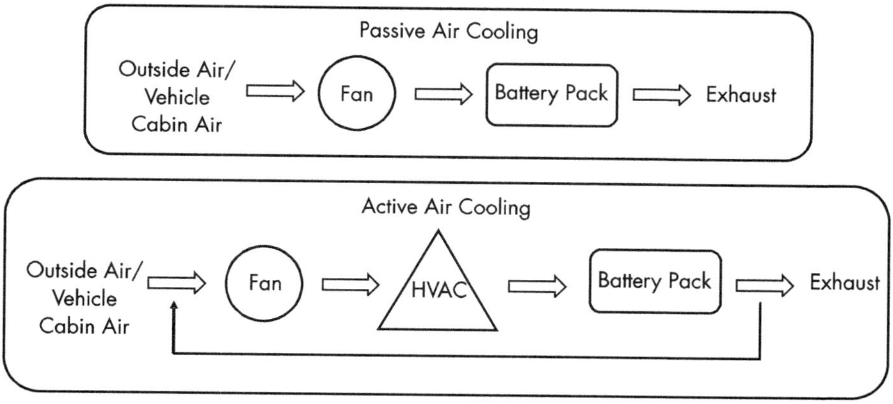


**Figure 3.5** Passive and active air cooling for an EV battery pack.

packs due to its relatively superior thermal dissipation capacity[7] as compared to alternative techniques (e.g., air cooling). However, it is also comparatively more costly and makes the battery pack heavier. In addition, coolant leakage inside the battery pack can be problematic. For this reason, EV battery packs incorporating liquid cooling systems require careful controls of the manufacturing process used for the coolant infrastructure as well as additional BMS requirements for detecting coolant leakage.

A third mode of cooling also used in some EV battery pack designs is a passive cooling technique that involves the use of phase change materials (PCM). As the name suggests, a PCM is a type of material that undergoes a phase change (e.g., from a solid to a liquid) as it absorbs heat. While absorbing the heat, the PCM stays at approximately the same temperature as the heat is absorbed as latent heat. The material reverts to its original phase once the heat is released or the temperature decreases. While the use of a PCM in EV battery pack designs is not widespread, some of its advantages such as the ease of maintenance, lower cost, and efficient heat dissipation and transfer make it an attractive option for thermal management in EV battery packs [11].

7. Thermal dissipation capacities vary considerably and are a function of the infrastructure implemented in the battery design.

## 3.3 BATTERY MODELING AND SOC DETERMINATION

Unlike Li-ion battery applications where cycling patterns and power demands are consistent and predictable such as in consumer electronic devices, cycling patterns and power demands in EV applications are much more dynamic and unpredictable. The charge and discharge of a Li-ion battery in an EV application will vary for a variety of reasons, such as an individual driver's charging patterns and driving behavior. The latter makes it much more difficult and costly to experimentally validate the performance of a Li-ion battery for an EV application.

One way to experimentally validate the performance of a Li-ion battery in an EV application is for the battery manufacturer to build a prototype and conduct trial-and-error testing to determine whether the battery will meet the application requirements. This testing will subject the battery to a wide range of real-world driving conditions and usage patterns. The test results can then be analyzed to determine what modifications should be made, and a new prototype can then be built. This process can be repeated until the prototype passes all required field tests. There are several drawbacks to this approach. Building a prototype battery is expensive and time-consuming. Additionally, because the number of prototype iterations that will be needed is unknown, accurately predicting the design cycle time within an acceptable window is very difficult. Determining what modifications will resolve identified issues and optimize battery performance will generally require a trial-and-error approach that is highly dependent on past experience with similar batteries. Matters can be further complicated as it is not atypical for additional performance requirements to be imposed later in the design cycle (e.g., by the EV manufacturer or government safety regulations), which may necessitate additional field testing and prototypes.

Battery modeling has thus become an essential tool for performance validation of batteries in EV applications. The use of an accurate battery model enables the validation of battery performance through computer simulations. This significantly reduces development time and costs. Simulation-driven design also allows for better performance optimization throughout the full range of real-world driving conditions, a more accurate prediction of the design cycle time, and a reduction in reliance on prototypes and field testing. For this reason, accurate Li-ion battery models are critical when selecting a battery for an application and determining whether the selected battery will meet the requirements for that application. For example, accurately characterizing the aging of a battery in an application, estimating the life cycle of the battery, determining whether the selected battery meets the power requirements for the application, and validating quick charge requirements all depend on robust battery models. A poor battery model for these simulation studies can provide misleading results, which can lead to costly delays in a battery development project.

In modern passenger EV applications, battery modeling is also important for the BMS as it enables the BMS to perform its functions more effectively. One of the most important quantities that is estimated with battery modeling is the battery's SOC. A battery's SOC provides the amount of energy available in the battery at a given time. Technically, SOC is defined as the percentage of the remaining charge inside a battery to the full charge and, thus, it can range from 0% to 100%. Inaccurate estimation of SOC can lead to misleading results regarding the suitability of a battery for an application and whether the battery meets performance requirements.

One method to estimate the battery SOC without using a battery model is coulomb counting. This method involves the integration of the measured current in and out of the battery. However, measurement errors, noise, and aging effects can lead to large errors in the SOC estimation using this model. This makes the use of a well-developed battery model essential for estimating battery SOC during operation.

### 3.3.1 Equivalent Circuit Model

The accuracy of SOC estimation using battery modeling is highly dependent on the quality of the battery model. There are two fundamentally different types of Li-ion battery models that are commonly used for battery SOC estimation: electrochemical model and equivalent circuit model. Electrochemical models attempt to simulate the internal physical and chemical properties of the battery, including the mass, energy, and momentum transport for each phase and component of the cell. The kinetics and thermodynamics of the chemical and electrochemical reactions are also typically included in the model. An electrochemical model is typically implemented as a system of complex coupled partial differential equations that must be solved in time for all spatial dimensions needed and, thus, have a large number of parameters that must be determined. The complexity and computational burden required to solve these complex equations make the electrochemical model less suitable for use in real-world EV applications.

The second type is the equivalent circuit model (ECM). In contrast to the electrochemical model, the ECM simulates the battery by using simple electrical components as an analog to the physiochemical principles. The use of these simple electrical components significantly reduces the complexity and computational burden and makes the ECM a suitable candidate for use in real-world EV applications. However, the accuracy of the ECM is highly dependent on the values used for the circuit parameters, which, in turn, depend on the amount and variety of experimental data collected.

The simplest ECM for a Li-ion cell represents the cell as an ideal voltage source (Figure 3.6). In this model, the voltage is not a function of current or past usage. This makes it unsuitable for any practical EV application as the cell's voltage will change depending upon a number of factors, including its SOC, temperature, the charge/discharge current, and other electrochemical reactions inside the cell such as slow diffusion processes.[8]

To account for these effects, additional electrical components can be added to the circuit, as shown in Figure 3.7. In this ECM, the cell's terminal voltage (V) is a function of the cell's open-circuit voltage (OCV), the cell's internal ohmic resistance ($R_0$)[9] and the diffusion effect in the cell represented by a parallel resistor ($R_1$) and capacitor ($C_1$) network.[10] The cell's OCV in this model is not constant but a function of the cell SOC. This model is often referred to as a first-order ECM.

---

8. When a cell is allowed to rest, its voltage does not immediately return to open-circuit voltage, but gradually decays, taking a long time to approach the open-circuit voltage. This is caused by slow diffusion processes in the cell.
9. This is a combination of the resistances arising from the active material of the anode and the cathode, the separator, and the electrolyte.
10. The time constant ($\tau$) of the parallel RC network represented by $R_1$ and $C_1$ is $R_1$ multiplied by $C_1$ (i.e., $\tau = R_1C_1$ and refers to the characteristic time it takes for the battery voltage to reach a certain fraction of its final value in response to a step change in current or load).

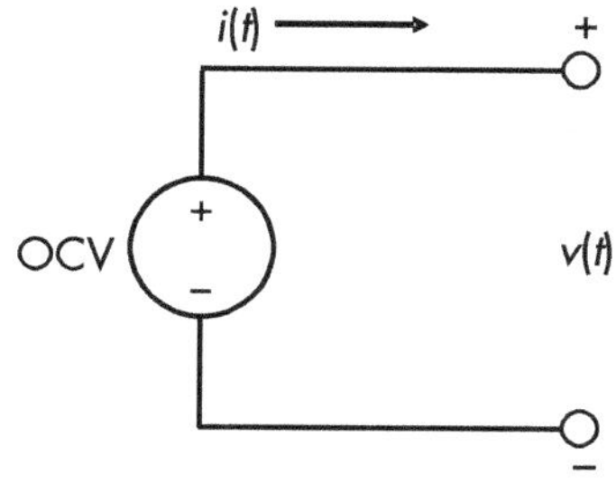


**Figure 3.6** The simplest ECM for a Li-ion cell.

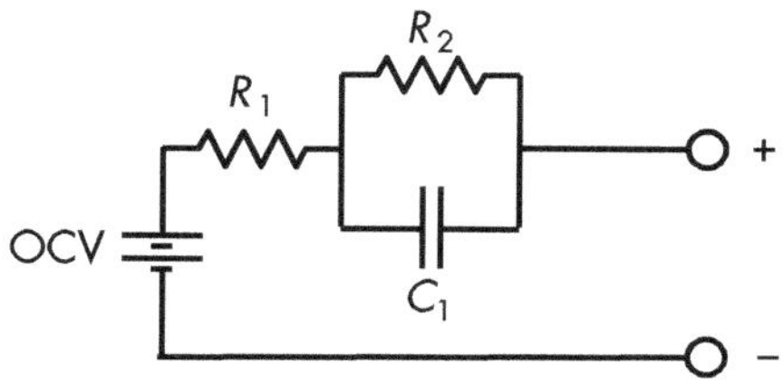


**Figure 3.7** First-order ECM of a Li-ion cell.

One or more additional RC pairs can be added to the first-order ECM to account for other conditions such as dynamic and instantaneous hysteresis [12, 13]. Figure 3.8 shows an example of a second-order ECM that is also commonly used during battery development. Although a second-order ECM model can provide improved accuracy as compared to a first-order ECM model, this improvement comes at the cost of additional complexity due to additional parameters that need to be determined. In addition, some of the accuracy gains are generally only observed at higher frequencies, which may not be needed for certain applications.

#### *3.3.1.1 ECM Parameter Estimation*

Once a battery cell ECM has been selected (such as the first-order ECM in Figure 3.7 or the second-order ECM in Figure 3.8), the model parameters need to be estimated (e.g., OCV as a function of SOC, $R_0$, $R_1$, and $C_1$ in Figure 3.7). Accurate SOC and other state estimations by the ECM require optimally calibrating the ECM parameters. For the ECM shown in Figure 3.7, the parameters that must calibrated include $R_0$ and the time constant $\tau = R_1C_1$.

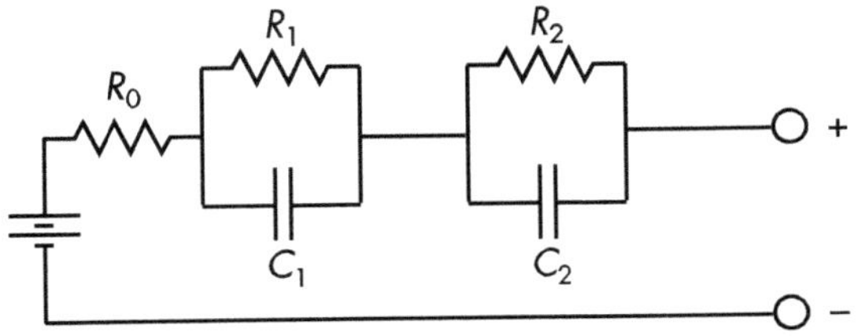


**Figure 3.8** Second-order ECM of a Li-ion cell.

Calibrated ECM parameters are typically provided in the form of parameter look-up tables that are generated during the development of the model. The look-up table for each ECM parameter contains values for different conditions such as temperature and SOC. For example, if the temperature is 20°C and the estimated SOC is 30%, the look-up table will have parameter values corresponding to those conditions that the ECM uses.

However, conventional parameter estimation and optimization techniques present a number of challenges and limitations. They require, for example, extensive experimental procedures and result in uncertainty as to whether the value of the estimated parameters is optimally calibrated. Some of the difficulties in estimating ECM parameters that accurately simulate the electrochemical and thermal dynamics of a battery are described in, for example, [14], which states that "when GA [genetic algorithm] is used in parameter identification, there are certain issues that cannot be avoided, such as the high computation time and easily falling into local optimum, i.e., local extremum."

Battery ECM parameter estimation generally requires capturing experimental data at different operating conditions such as ambient temperature, SOC ranges, current, and battery age. Various parameter estimation and optimization techniques are described in the literature, including indirect estimation and parameter optimization, numerical searching methods, conventional fitting methods, and global optimum searching methods. These techniques use data from a variety of tests and other related experimental data (such as hybrid pulse power characterization (HPPC) test data) to estimate and optimize the values of the ECM parameters [13, 15].

When using look-up tables, the parameter values for intermediate conditions need to be estimated, and linear interpolation is a common way of estimating these intermediate parameter values. However, linear interpolation assumes relative linearity between the look-up table values for adjacent conditions. While this assumption of linearity may be valid for some conditions, it does not hold for many corner cases such as when the battery is at a low SOC and/or temperature. Conventional methods that utilize linear interpolation for intermediate parameter values perform some smoothing of parameters to account for these boundary conditions and to improve the accuracy of the model.

#### *3.3.1.2 ECM Parameter Estimate Validation*

Once the parameters for a chosen ECM have been identified, extensive validation is required to ensure that the model will provide a reliable estimate under different operating conditions such as ambient temperature, SOC ranges, current, and battery age. ECMs are typically validated by comparing simulated voltage as determined by the model to experimentally measured voltages or by using experimental data to validate the model's SOC, cell center temperatures, and terminal voltage. ECMs can also be validated using discharge-charge experimental tests and microstructural data. Often a separate data set (called a validation data set) is used for evaluating the performance of the model. This data set should ideally cover a wide range of operating conditions to ensure the model's robustness. This validation data set is used to assess the accuracy and reliability of the ECM and involves comparing the model's prediction with the actual measurements from the validation data set. Various performance metrics are used to quantify the goodness-of-fit between the model predictions and the validation data set. These metrics may include root mean square error (RMSE), mean absolute error (MAE), and coefficient of determination (R-squared). If the model fails to meet the desired accuracy criteria, further refinements may be made by adjusting the model structure or fine-tuning the parameters. This iterative process continues until satisfactory agreement between the model predictions and the experimental data is achieved. Any uncertainty associated with the model predictions is determined through sensitivity analysis or Monte Carlo simulations to understand how variations in the model's input parameters can affect the model output.

## 3.4 RELIABILITY, SUSTAINABILITY, AND END-OF-LIFE RECYCLING REQUIREMENTS FOR EV BATTERIES

A significant issue to tackle with EVs involves the challenge of developing an effective and environmentally friendly battery recycling infrastructure. Since the EV battery packs require critical metals including lithium, cobalt, nickel, and copper, various governmental entities around the world have recognized the need for regulated recycling of these materials to help meet demand, prevent depletion of mineral reserves, and reduce environmental impacts [16]. The latter effectively targets creating a close loop where, once batteries reach the end of their life cycle, these resources are extracted and reintroduced into the manufacturing process, thus cutting down on mining costs and mining-related pollution and preventing hazardous materials from entering the waste stream to achieve a sustainable and reliable battery production cycle [17].

EV battery recyclers usually target retrieving metals in the cathode, such as cobalt and nickel, as they are expensive and thus economically worthwhile [18]. Today, there are three main recycling techniques for EV batteries:

- Direct recycling;
- Pyrometallurgy;
- Hydrometallurgy.

Direct recycling [19] involves disassembling the battery packs and extracting the anode and cathode materials from the cells for reconditioning and reuse. During a direct recycling process the lithium content within the active material is replenished through the incorporation of new cathode materials such as $Li_2CO_3$ or $LiOH/Li_2SO_4$. Direct recycling is ideal due to the quantity of materials that are recovered after the recycling process as well as the fact that it leaves the cathode morphology mostly intact. However, direct recycling also faces challenges since disassembling an EV battery pack cleanly can be difficult, costly, and time-consuming due to the presence of potting materials, hardware, and adhesives used for structural support, electrical connectivity, and passive propagation resistance infrastructure. Consequently, production costs tend to be higher when compared to other recycling methods.

To address the difficulty in the disassembly of these battery packs, some battery recyclers resort to shredding spent batteries. The shredding process [20] results in few materials, the most important being the shredded anode and cathode active material known as "black mass." The two common processes used to extract metals, such as nickel, cobalt, and lithium, from the black mass following the shredding process are pyrometallurgy and hydrometallurgy. Pyrometallurgy consists of burning (smelting) the spent active material, after which various techniques can be used to extract the metals. Hydrometallurgy employs acids to leach the metals out of the spent active material. After leaching or smelting, the newly extracted materials need to be processed further to be made into new batteries, at this point, following the same procedure for battery manufacturing used for nonrecycled materials. Though these two processes prove advantageous due to their relative ease and lower production costs, they lack the quality and quantity of the recovered materials as compared to the direct recycling method [19].

Regarding regulatory efforts within the United States to date, no federal or state mandates directly address best practices or guidance for Li-ion battery recycling used in EV applications [21]. As described by the National Renewable Energy Laboratory (NREL) in 2021 [22], "it is unclear how decommissioned lithium-ion batteries are legally defined in terms of waste." In an attempt to address this confusion, the Environmental Protection Agency (EPA) issued a memorandum in May 2023 clarifying how the hazardous waste regulations for universal waste in Title 40 of the Code of Federal Regulations (CFR) part 273 apply to the recycling of Li-ion batteries [23]. In the EU, Regulation (EU) 2023/1542 concerning batteries and waste batteries was put into effect in August 2023. This regulation not only addresses recycling requirements but sets down the requirements for sustainable, reliable, and diligent battery use from the point the batteries are manufactured to their end of life. Additionally, this regulation places extended responsibility and accountability on manufacturers for the collection, treatment, and reporting of spent batteries.

## 3.5 PROPAGATING CIRCUIT BOARD FAILURES

A typical EV battery pack will contain one or more printed circuit boards (PCBs) that monitor and control the relatively high battery voltages (several hundred volts) and are surrounded by an infrastructure that often utilizes an electrically conductive coolant. An introduction of any coolant onto the PCB can increase the risk of a circuit board failure, which can be a trigger for a larger battery fire, as mentioned previously in this chapter. However, PCB failures can occur due to a number of reasons and not just due to the introduction of coolant. These include:

- The design or manufacturing of the PCB itself;
- Conduction between components or traces on the PCB that were designed to be electrically isolated;
- A component on the PCB;
- A component whose failure affects surrounding components;
- A component that results in the propagation of the failure.

Propagating circuit board failures are especially problematic in EV batteries as they can heat up the cells and initiate a cascading event that leads to a larger battery fire. The initiating mechanism for a propagating PCB failure is the formation of a resistive path between two traces or layers at different potentials. This resistive path can form due to an external heat source, insulation breakdown, poor construction/soldering techniques, arcing, or contamination. A propagating PCB failure typically starts slowly from an initial resistive path between two conductors at different electric potentials. The heat dissipated through the resistive path pyrolyzes the insulating material. The pyrolyzed insulating material is electrically a poor conductor and adds to the heat generated by the original resistive fault. The resulting charred insulating material is resistive and continues to dissipate heat, pyrolyzing more insulating material around it. When the charred resistive area becomes large enough, the propagating PCB fault accelerates and continues to move toward the source of the power feeding the fault. During this process, it is possible that electrical arcs form between the different electrical potentials across the charred material and participate in propagating the fault. The propagating fault terminates when the source of the electric potential difference is removed or the fault runs out of PCB material (Figure 3.9).

Common circuit protection devices (e.g., fuses and PTCs) do not always terminate a propagating circuit board failure because the resistive nature of the fault is dynamic and usually limits the current in the fault to less than the activation threshold of the protection device. Understanding PCB areas that are susceptible to propagating circuit board failures is important in EV batteries to ensure that these PCB areas have adequate protection against propagating circuit board failures. PCB areas typically susceptible to propagating circuit board failures include:

- Areas on battery monitoring boards where the cell voltage sense lines that provide information on individual cell voltages are connected;
- Areas on the central BMS where the battery pack voltage is connected for monitoring;
- Areas on any PCBs where a single point of failure can lead to resistive heating.

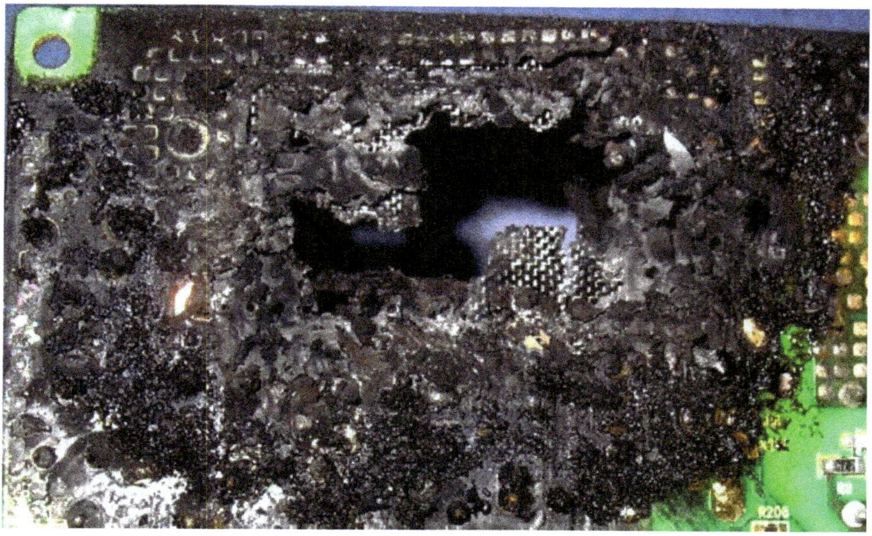

**Figure 3.9** An example of a propagating circuit board failure.

## References


[1] International Organization for Standardization (ISO), "Road Vehicles - Functional Safety," ISO 26262:2018, 2018, https://www.iso.org/home.html.

[2] Mendias, M., A. Arora, and S. Lele, "Functional Safety & Safety Critical Systems - An Overview," SAE Technical Paper 2021-01-0157, 2021.

[3] International Electrotechnical Commission, "Functional Safety of Electrical/Electronic/Programmable Electronic Safety-Related Systems," IEC 61508:2010, 2019, https://webstore.iec.ch/.

[4] Arora, A., and S. Lele, "Functional Safety & Li-ion Batteries," *Battcon*, Orlando, FL, 2019.

[5] Imre, T., C. Buidin, and F. Mariasiu, "Battery Thermal Management Systems: Current Status and Design Approach of Cooling Technologies," *Energies*, Vol. 14, No. 16, 2021, p. 4879.

[6] Garud, K. S., et al., "A Review of Advanced Cooling Strategies for Battery Thermal Management Systems in Electric Vehicles," *Symmetry*, Vol. 15, No. 7, 2023, p. 1322.

[7] Fu, P., et al., "A Review of Cooling Technologies in Lithium-Ion Power Battery Thermal Management Systems for New Energy Vehicles," *Processes*, Vol. 11, 2023, p. 3450.

[8] https://ev-lectron.com/blogs/blog/ev-battery-cooling-how-does-it-work.

[9] https://www.laserax.com/blog/ev-battery-cooling.

[10] https://evmagazine.com/technology/ev-battery-cooling-why-is-it-so-important.

[11] Grimonia, E., et al., "Thermal Management System Using Phase Change Material for Lithium-Ion Battery," *Journal of Physics*, Conf. Ser. 2117 012005, November 1, 2011.

[12] Hu, Y., et al., "Electro-Thermal Battery Model Identification for Automotive Applications," *Journal of Power Sources*, Vol. 196, 2011, pp. 449–457.

[13] Plett, G., "Extended Kalman Filtering for Battery Management Systems of LiPB-Based HEV Battery Packs Part 2. Modeling and Identification," *Journal of Power Sources*, Vol. 134, 2004, pp. 262–276.

[14] Kai, H., et al., "Development of Accurate Lithium-Ion Battery Model Based on Adaptive Random Disturbance PSO Algorithm," *Mathematical Problems in Engineering*, Vol. 2018, July 3, 2018.

[15] Rothenberger, M., et al., "Genetic Optimization and Experimental Validation of a Test Cycle That Maximizes Parameter Identifiability for a Li-Ion Equivalent Circuit-Model," *Journal of Energy Storage*, Vol. 4, December 2015, pp. 156–166.

[16] Baum, Z. J., et al., "Lithium-Ion Battery Recycling—Overview of Techniques and Trends," *ACS Energy Letters*, Vol. 7, No. 2, 2002, pp. 712–719.

[17] https://afdc.energy.gov/vehicles/electric_batteries.html.

[18] https://www.science.org/content/article/millions-electric-cars-are-coming-what-happens-all-dead-batteries.

[19] Harper, G., et al., "Recycling Lithium-Ion Batteries from Electric Vehicles," *Nature*, Vol. 575, 2019, pp. 75–86.

[20] https://www.epa.gov/hw/lithium-ion-battery-recycling.

[21] Bird, R., et al., "The Regulatory Environment for Lithium-Ion Battery Recycling," *ACS Energy Letters*, Vol. 7, No. 2, 2022, pp. 736–740.

[22] https://www.nrel.gov/news/program/2021/pathways-to-achieve-new-circular-vision-for-lithium-ion-batteries.html.

[23] United States Environmental Protection Agency, "Lithium Battery Recycling Regulatory Status and Frequently Asked Questions," 2023.

# 4

# CHARGERS

## 4.1 CHARGER TYPES

An often-discussed challenge in increasing the adoption rate of EVs involves EV battery chargers. Considerable efforts are being put into both achieving faster charging times (i.e., higher power-delivering capabilities) for EV batteries and providing widespread access to charging stations by expanding the infrastructure. Charging times for EV batteries while steadily reducing with time still remain considerably longer than the time that it takes to fill the gas tank of a conventional internal combustion engine (ICE) vehicle. The deployment of EVs will depend on the driving range, performance, costs of batteries, and the convenience of recharging the batteries. Researchers contend that the charging time of EV batteries needs to get much closer to the time required to refuel ICE vehicles and that EV supply equipment (EVSE) needs to extend with higher power levels for ultrafast charging. This will require smart charging with coordinated control techniques, and higher-power converters that can reduce the charging time. Extensive research is currently being conducted on establishing advanced control strategies, architectures and converter topologies for EV charging systems [1].

Li-ion EV cells are charged using a specific constant current-constant voltage (CCCV) charge profile as shown in Figure 4.1. The charge profile consists of two distinct phases as denoted by the name: a constant current (CC) phase and a constant voltage (CV) phase. During

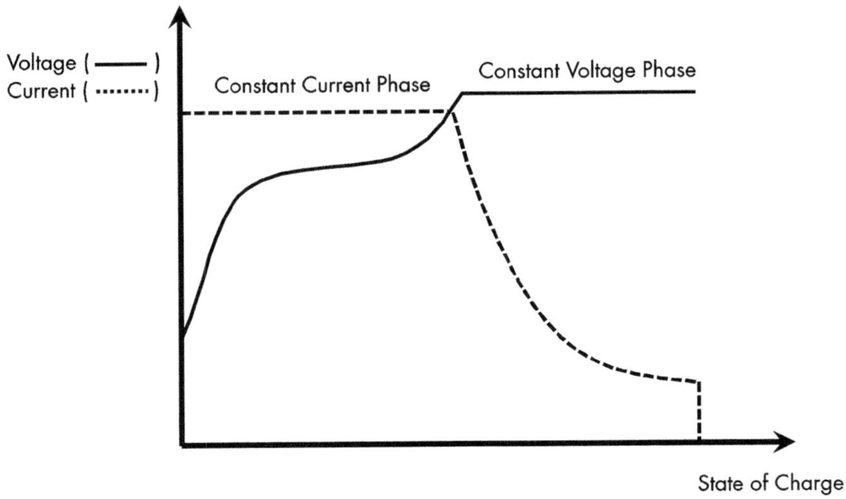


**Figure 4.1** Constant-current-constant voltage (CCCV) charging profile.

the CC phase, the charging current is held constant until the desired cutoff voltage is achieved.[1] The magnitude of the charge current during the CC phase is defined by the cell manufacturer and typically varies with ambient temperature. Typical charging currents specified by the cell manufacturer at room temperature are between 0.3C to 1C.[2] Once the cutoff voltage is reached, the charger transitions to the CV phase, during which the voltage is held constant, resulting in a drop in the current. Charging is terminated once the current drops to a cutoff value (typically the current decreases to a certain current cutoff of approximately 0.1C or lower).

A variety of methods can be used for charging EV batteries. Figure 4.2 shows a high-level classification of the different EV charging methods currently in use globally. By far, the most common method of charging an EV battery is conductive charging, which takes advantage of simplicity and relatively low cost since all that is required is a conductive cable, sockets, plugs, and electrical contact between the charger and the EV battery. Conductive charging can be achieved through

1. The cutoff voltage to which each battery cell is charged depends upon the exact Li-ion cell chemistry.
2. The C-rate of a battery is calculated as the battery's capacity in Amp-hours (Ah) divided by hours. For example, a cell rated with a capacity of 100 Ah will have a 1C rate of 100A, 0.5C rate of 50A, and so on.

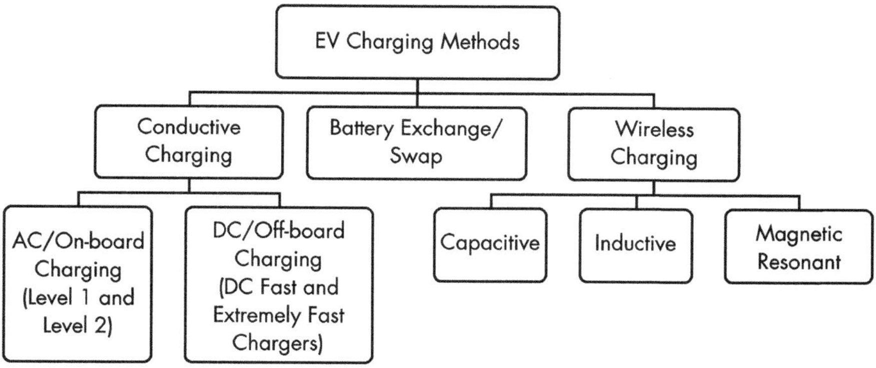


**Figure 4.2** EV charging methods.

both AC and DC charging. Other charging methods include wireless charging, which was first introduced several years ago and has started to gain popularity with both automobile manufacturers and standards developing organizations such as the Society of Automotive Engineers (SAE) International. Battery swapping, which involves driving the vehicle to a designated location and swapping the battery with a fully charged battery, is also becoming popular, especially in parts of Asia. As an example, companies such as NIO Power [2] advertise swaps to fully charged batteries within approximately 3 minutes.

### 4.1.1 Conductive Charging: AC and DC Chargers

As shown in Figure 4.1, EV batteries require a specific direct current (DC) charge profile for charging. This necessitates both regulation and monitoring of the DC power delivered to the battery. This DC power can be provided through either an alternating current (AC) charging station or a DC charging station.

An AC charging station[3] acquires the power from the grid or other renewable sources and efficiently transfers this power to the EV's on-board charger (OBC). The OBC then executes the rectification and power conversion of the AC power, thus converting the grid power into DC power that can be used to charge the battery. During this process, the charge controller within the AC charging station and the OBC within the EV are in constant communication and monitor

3. The terms charging station and chargers are used interchangeably in this book.

the charging process including the temperature of the charging cable and the connection status of the charging plug.

A DC charging station also acquires power from the grid or renewable sources but internally performs both power rectification and conversion within its off-board charger and provides the converted DC power directly to the EV battery. The AC-DC conversion by the DC charging station removes the need for the OBC in the charging process and the power supplied by the DC charging station can directly charge the battery. A DC charging station has a relatively more complex architecture when compared to an AC charging station due to its ability for internal power rectification and conversion. Just like the AC charging station, the DC charging station also constantly communicates with the EV battery during the charging process.

Table 4.1 summarizes the charging levels and power delivery capabilities of commonly installed AC and DC charging stations [3–6].

In addition to the amount of power that can be delivered, an additional difference between AC and DC chargers is the rate at which this power is delivered. AC chargers typically have a relatively constant power delivery rate regardless of the battery SOC. In contrast, the power delivery rate for a DC charger is dependent on the battery SOC and ramps down once the battery SOC reaches approximately 80% resulting in the time taken to charge the battery from 80% SOC to 100% SOC taking proportionally much longer as compared to the time taken to charge to the first 80% SOC.

As will be discussed in Chapter 8, EV charging connectors, inlets, and plugs used in conductive charging are designed and manufactured

**Table 4.1**
EV Charger Levels and Power Delivery Capabilities

| **Charger Type** | **AC Level 1** | **AC Level 2** | **DC Fast and Extremely Fast Charger** |
|---|---|---|---|
| EV Connector | SAE J1772 | SAE J1772 | CHAdeMO |
| | | GB/T (AC) | CCS |
| | | Mennekes | GB/T (DC) |
| | | NACS (Tesla) | NACS (Tesla) |
| Power (kW) | 1–2 | 3–22 | 50–400 |
| Estimated range per hour of charging (miles) | 3–5 | 10–25 | 100–780+ |

*From:* [3–6].

per standards such as SAE J1772, IEC 62196, and GB/T 20234. These standards provide requirements for safety, environmental robustness, locking and latching mechanisms, mechanical performance, electrical performance, as well as overall guidelines for the design of the charging couplers. The charge coupler (connector) design depends on the power delivery capabilities of the charging station. However, the main pins designation embedded within these connectors are mostly consistent across the various connector designs. This is to ensure uniformity and ease for the user regardless of their EV make and model. The charge coupler connector pins and their functions are as follows:

- *Proximity pilot (PP)* or proximity detection[4] is a communication line that is used to ensure the proper connection between the charger and the EV through the use of specifically rated resistors as shown in Figure 4.3 [7].
    - The charging station detects that it is connected to an EV through a switch on the handle of the charging connector. When the switch is pressed, the PP signal detects a change in the cable resistance, which is used as a signal for the disconnection between the charging station and the EV. When a disconnection is detected through the PP signal, the charg-

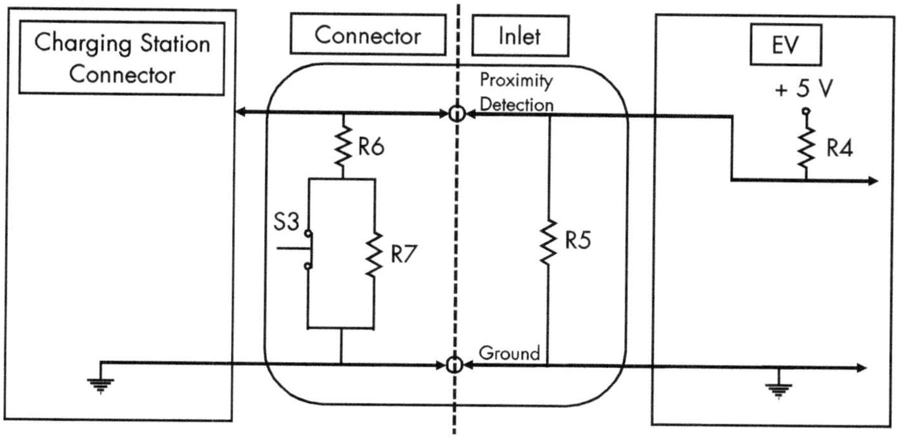


**Figure 4.3** Proximity pilot equivalent circuit. (*After:* [7].)

4. Also sometimes called the plug detection signal.

ing current is stopped immediately to mitigate any electrical arcing-related hazard as the charging connector is removed.

  - The PP signal is also used in some charging station designs as an indication of the charging cable's current carrying capacity. This is specifically relevant for charging stations not utilizing permanently connected charging cables. The EV interrupts the current supply if the current carrying capability of the cable is exceeded as detected by the measurement of the in-line resistor.

- *Control pilot (CP)* is a communication line used to monitor the charging process (Figure 4.4 [7]). The CP signal is a 1-kHz square wave at ±12V generated by the charging station to ensure that a connection has been established between the charging station and the EV, to check for grounding continuity and to determine the maximum allowable current that the EV can draw during the charge process. The CP line is also used to control the start and end of the charging process.[5]

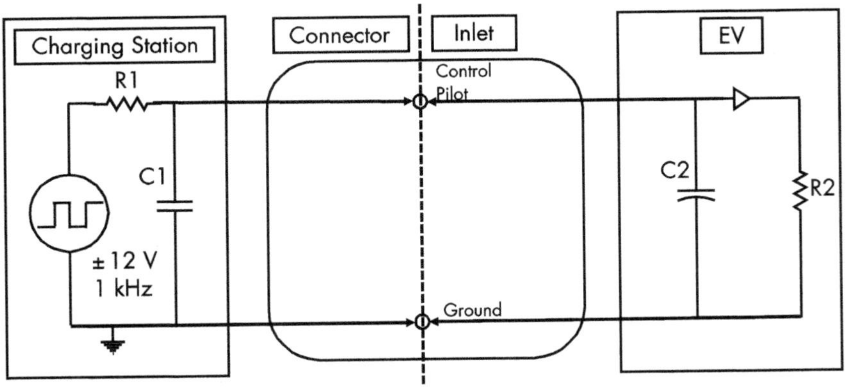


**Figure 4.4** Control pilot equivalent circuit. (*After:* [7].)

5. For certain charging connectors such as CHAdeMO and BB as defined in IEC 62196-3, the CP signal is a CAN signal (C-H/S+: CAN High and C-L/S–: CAN Low). CAN stands for controller area network and is a serial bus system that allows communication between intelligent devices. CAN-based systems are common within the automotive industry as they allow centralized communications between various vehicle components without the need for extensive copper wiring.

- *Protective Earth (PE) or Ground (G)* is the ground connection for the charging station and the EV chassis.
- *Main Power Pins (L1, L2, L3, N, DC+, DC−)* are used as the current carrying pins for power transfer between the charging station and the EV.
  - As discussed, charging stations are capable of accommodating AC or DC charging with some stations capable of doing both through the use of the Combined Charging System (CCS) protocol. These charging connectors contain pins for both AC charging (L1, L2, L3, N) and DC charging (DC+ and DC−).[6]

The charge cycle requires a very specific sequence of signaling between the charger and the EV before charging is enabled. This sequence is established through the construction of the charger and EV charge connectors, which results in the different connector pins making electrical contact at different times as the charger is connected to the EV. As an example, per SAE J1772, for an AC charging station, the following sequence is used by the signaling circuit [7, 8]:

1. *Charging station determines connection with EV:* As the charge connector is embedded into the vehicle inlet, ground continuity is first established between the charging station and the EV. The main current-carrying pins are then connected, though no current is yet flowing, followed by the PP and CP pins connection. As the connector latch is released, the locking mechanism is enabled, which triggers the PP signal. Once the charging cable is securely latched, the CP signal is enabled indicating the charging station's readiness to supply power.
2. *Ventilation requirements and EV readiness are determined:* If ventilation is required throughout the charging process, the charging station will also need to power a ventilation system. Additionally, if these requirements are not properly met, the charging sequence will not be able to commence.

6. The BB connector defined in IEC 62196-3 and GB/T 20234-3 also contain low-voltage auxiliary DC power pins A+ and A−.

3. *Charge initiation:* Once the charge readiness is determined for both the charging station and the EV, the charging station proceeds to close contactors and supply current.
4. *Charge current adjustment and ground continuity monitoring:* Once charging commences, the charging station and EV are in constant communication to ensure ground continuity and proper power delivery.
5. *Charge termination:* If the connector latch is released during the charging process, the charging station determines a change in the PP signal and terminates the charging current. The CP signal determines that the EV is not ready to charge and the charging station opens the contactors terminating power delivery to the EV. When removing the connector, the CP and PP pins are first disengaged, followed by the main current-carrying pins, and, lastly, the ground is disconnected.

### 4.1.2 Battery Swapping

One of the advantages, from a consumer's perspective of traditional ICE vehicles is the speed at which the fuel can be replenished. Filling up a vehicle with gasoline will usually take a few minutes as compared to charging an EV battery, which takes from tens of minutes to a few hours even when using the fastest chargers available in the market today. The relatively slower charging speeds have given rise to the battery swapping [9] industry in parts of Asia, Europe, and, more recently, in the United States, which aims at matching the speed and convenience of visiting a gas station.

Succinctly, battery swapping is a method by which a depleted battery is swapped with a fully charged battery at dedicated locations that have the means to do this process efficiently, precisely, and safely. This technology also aims to tackle additional concerns currently present when using standard charging stations. When using conventional AC or DC charging stations, the demand on the grid can be substantial and unpredictable. The infrastructure in various regions globally is not yet fully developed to accommodate a large portion of the population owning and charging their EVs. Battery swapping eases this issue as it allows for the charging of the depleted battery to happen in a more controlled manner. Additionally, the swapped batteries often do not need to be charged very rapidly or right away, allowing them to be charged using a more modest charger as opposed to a fast charger. This has the added benefit of extending the lifetime of the battery due to the lower charging speeds [9].

Nio Power has emerged in China as a leader in battery swap technology and has also expanded to parts of Europe. Within the United States, various start-ups, such as Ample [10], have invested in this technology. However, battery swapping technology also faces challenges due to several reasons. These include the relatively more expensive infrastructure for a battery swap station when compared to an AC or DC charging station and the complication of the unique battery pack chemistries and designs of each EV manufacturer. Due to the cost accrued in building up the infrastructure, various battery swap companies plan to charge customers a monthly fee for the services. Though the technology is expanding and growing, further research and development are needed for it to be adopted as a primary method of EV charging.

### 4.1.3 Wireless Charging

Wireless charging of an EV is accomplished through the transfer of energy between a transmitter and a receiver via an electromagnetic field. The typical design infrastructure of a wireless charging station is shown in Figure 4.5 [11, 12].

As with AC and DC charging stations, a wireless charging station acquires power from the grid that needs to be rectified and converted to be able to be received by the battery. The rectification and conversion process occurs within the charging station itself. Following the latter, a high-frequency inverter is used and impedance matched before it can be received by the transmitter pad. The high-frequency inverter is used to convert the rectified DC power to a high-frequency AC signal that can be used by the transmitter. The transmitter then couples to the receiver, which is located within the EV to enable the power transfer. Once the power is transferred to the receiver pad, it is again impedance matched before it can be received by the OBC[7] and ultimately delivered to the EV battery.

The main coupling modes between the transmitter and receiver can be inductive, capacitive, or magnetic resonant [13–16]:

7. A detailed discussion on OBCs is provided in Section 4.2.

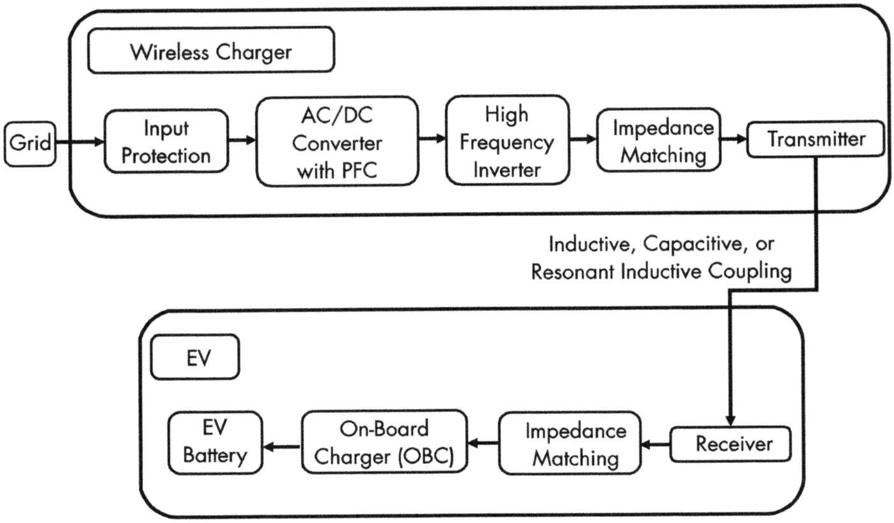


**Figure 4.5** Wireless charger infrastructure.

- *Inductive coupling* relies on the principle of electromagnetic induction where energy is transferred using rapidly fluctuating electric and magnetic fields between two physically separated conductive coils. For an EV wireless charger, the high-frequency inverter supplies the transmitter coil with the rapidly fluctuating electric field that it requires to, in turn, create a fluctuating magnetic field per Faraday's law of induction. This rapidly fluctuating magnetic field is then coupled to the receiver coil, which, in turn, generates a fluctuating electric field allowing a current to flow through it and thus permit the transfer of power between the two coils.
- *Capacitive coupling* uses two conductive capacitor plates, one acting as the transmitter and the other as the receiver, to transfer power through a static electric field. The limiting factors for maximum power transfer using this methodology are the size and distance between the capacitive plates.
- *Magnetic resonant coupling* is similar to inductive coupling with the addition of capacitors in series with the transmitter and receiver coils. This method has the added benefit of a lon-

ger transmission distance as compared to inductive coupling with the drawback being that the resonant frequency is difficult to adjust.

Although EV wireless charging is still considered an up-and-coming technology, various advantages have been identified when compared to standard conductive charging. With wireless charging, no wires are needed to plug into the EV eliminating any shock or burn hazards associated with a charging coupler fault. Wireless charging also promises a reduction in charging times as well as in emissions. However, as with all technologies, wireless charging also faces certain challenges [17], including electromagnetic interference (EMI) creation, heat production during the charge cycle, close proximity required between the transmitter and receiver, power transfer efficiency, alignment tolerances between transmitter and emitter plates, and the need for the implementation of a completely new charging infrastructure.

## 4.2 CONDUCTIVE CHARGER DESIGN ARCHITECTURES

Evaluation of an EV charging technology requires understanding the charging method, the power flow direction, the location of the charger (whether on-board or off-board) and the power supply technique. As already discussed, the basic unit of an EV charging system is the EV supply equipment (EVSE) which acts as an interface and connects the local electrical supply (i.e., grid) to the EV. The design of an EVSE necessitates the integration of several high-power loads and an adaptation to the existing electrical infrastructure. The EV charging station as defined by IEC 61851-1, is the stationary part of the EVSE that is connected to the supply equipment. It can either be wall-mounted or floor-standing, AC or DC.

As discussed previously, EV charging methods can be conductive, wireless or involve the swapping of batteries. Since battery-swapping and wireless charging are still at early stages for large-scale implementation, this section will discuss conductive charging. Both on-board and off-board charger implementations use a wired connection for the charging process.

### 4.2.1 On-Board Chargers

On-board chargers (OBCs) take power from the AC grid and condition this power to provide the right voltage and current to charge the battery. Due to space limitations, OBCs need to be designed to support a high-power density, high efficiency, and good thermal management for heat dissipation. These requirements limit the power output that can be achieved using OBCs. Vehicles with OBCs connect to a standard single-phase or three-phase AC charging station with all the power rectification and conversion performed by the OBC. Figure 4.6 [18] shows a high-level topology of an example OBC that is designed with an interleaved AC/DC converter with power factor correction (PFC),[8] followed by a pulse-width modulation (PWM) full-bridge DC-DC converter [18].

Active power factor conversion (PFC) in OBCs is typically performed using a boost converter. In a typical design, the PFC control circuit periodically shorts an inductor to ground causing energy to build up in the inductor. When the shorting switch is opened, the energy built up in the inductor is directed to a capacitor for storage. Since the inductor acts as a current source in series with the input, the output voltage ends up being higher than the input voltage (e.g., 170V for a 120 VAC input or 340V for a 220–240 VAC input). In this design, the output is not isolated from the input. For this reason, this design in an OBC application typically incorporates an additional switch mode converter after the boost PFC stage to provide isolation

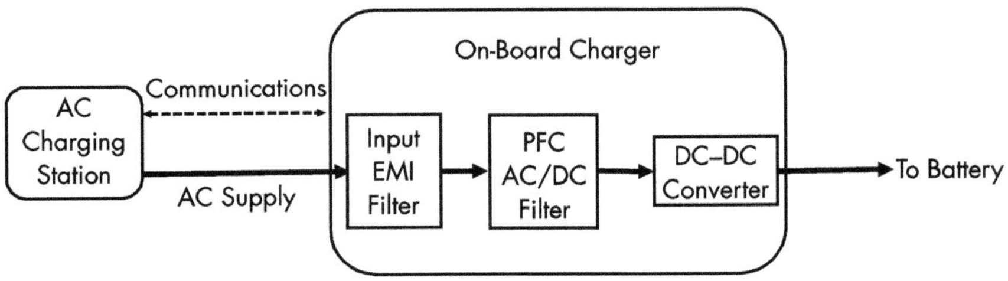


**Figure 4.6** On-board charger architecture.

8. Power factor is the ratio between the useful (true) power (kW) to the total (apparent) power (kVA). Power factor correction is the term given to the technology to restore the power factor to as close to unity as possible. This is often achieved by the addition of capacitors to the electrical network that compensate for the reactive power demand of an inductive load.

and tighter regulation. Figure 4.7 shows a block diagram of an active PFC using a boost converter [19].

As previously discussed, when the charging station and the EV are first connected, the charge controller in the charging station communicates with the EV. This communication includes information related to connectivity, any fault conditions, and the current limits for the OBC. Once communications are established, power flows from the AC charger to the OBC to charge the battery. Typical components in the OBC include:

- An EMI filter at the input that is typically required to ensure compliance with EMC standards. The EMI filter reduces both differential-mode and common-mode noise and suppresses the noise emitted by the charger that may be introduced back into the grid.
- The DC-DC converter takes the output from the PFC and converts it into a dedicated DC output that follows the required battery charging profile. There are many different possible DC-DC converter designs that can be used in an OBC. A phase-shifted full-bridge (PSFB) topology with zero-volt shifting (ZVS)[9] is a commonly used design due to its high efficiency and lower EMI. A PFSB converter consists of four power electronic switches (e.g., MOSFETs) that form a full-bridge on the primary side of an isolation transformer and diode rectifiers or MOSFET switches for synchronous rectification on the secondary side. ZVS occurs when the MOSFETs are turned off

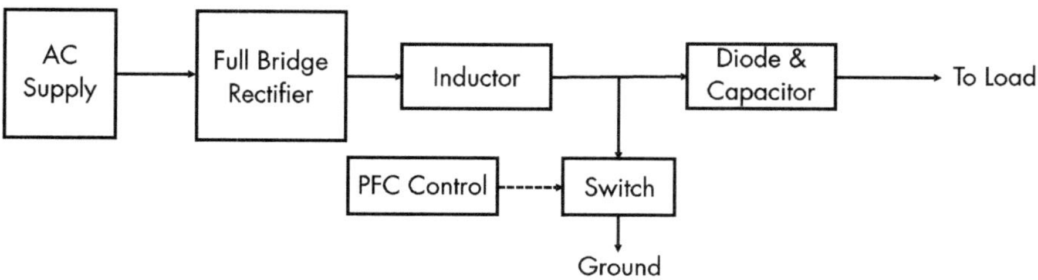


**Figure 4.7** Active PFC using a boost converter.

9. ZVS is a technique that uses a PWM-based operation but with an additional phase to the PWM timing to allow for ZVS operation. ZVS enables the voltage regulator to engage soft switching, which avoids switch losses that are typically incurred during conventional PWM operation and timing.

and on when their respective drain-to-source voltages are at or near 0V, effectively eliminating the instantaneous turn-on power loss of the MOSFETs caused by the drain-to-source capacitance [20].

- A microcontroller that monitors the overall operation of the system and generates the output voltage and current references to manipulate the voltage and current loop control of the DC-DC converter. It also typically monitors temperature, controls the cooling circuit, interacts with any user interfaces and communicates with other systems in the vehicle.

### 4.2.2 Off-Board Chargers

The architecture of off-board EV charging stations can be AC bus-based, DC bus-based, or a combination of AC and DC bus structures. Figure 4.8 [1] shows the high-level architecture of AC bus-based and DC bus-based charging stations. In an AC bus-based system, the EV charging station requires an AC-DC rectifier and a DC-DC converter. This increases the cost, complexity, and power conversion stages and reduces the efficiency of the charging station. In contrast, DC bus-based systems utilize a single AC-DC converter on the grid side to provide DC power to the common bus with each charging station only containing a DC-DC converter. This leads to a more efficient, cost-effective, smaller, and flexible charging station design as compared to a charging station in an AC bus-based architecture.

A survey of ultrafast charging systems today will show that a majority of these systems are designed as AC-bus based charging systems that use well-equipped configurations and a mature power converter infrastructure. Added advantages that make them attractive include their ability to allow for the direct usage of local loads, a capability not afforded by DC power-based fast charging stations. However, the industry is slowly shifting to DC power-based fast charging stations as these stations afford a relatively higher efficiency, can be manufactured and operated at comparatively lower costs, and provide additional flexibility in integration with the distribution grid when compared to AC-bus based charging systems.

The off-board AC bus-based charging stations consist of two converter stages: an AC-DC converter, and a DC-DC converter. The AC-DC

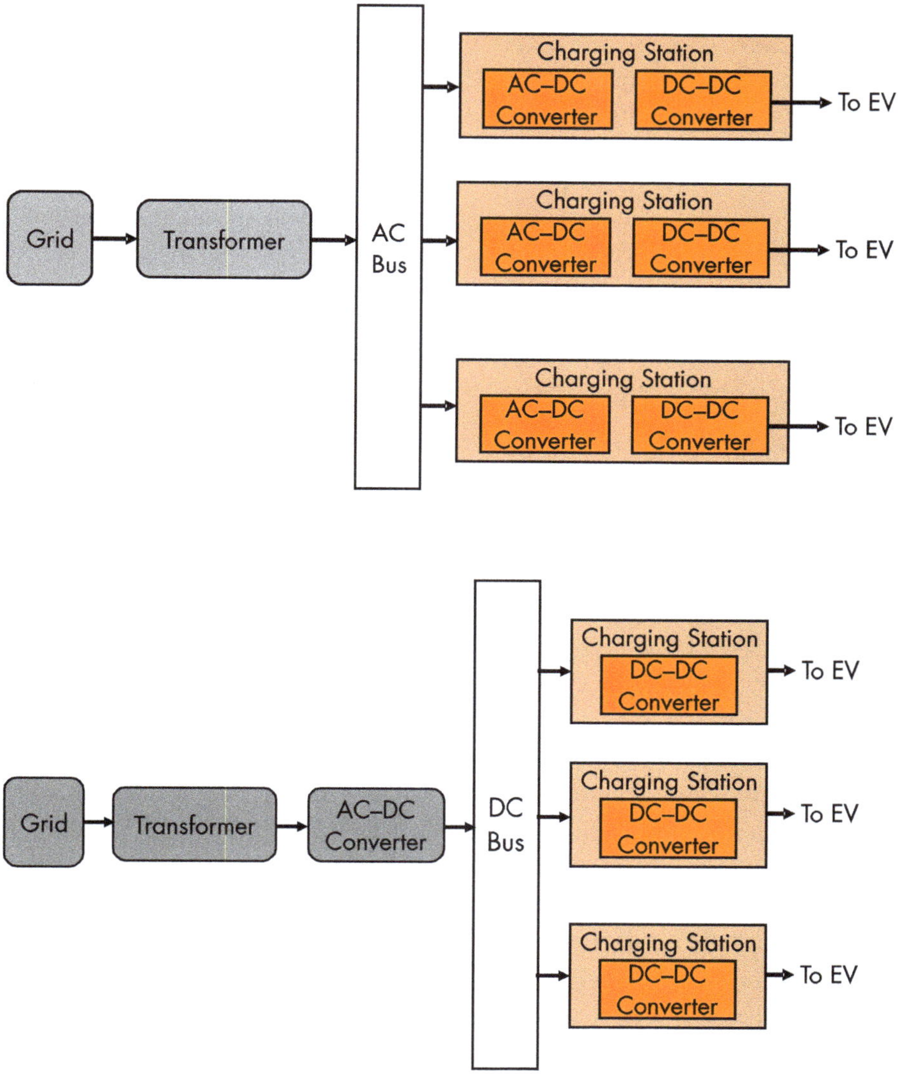


**Figure 4.8** AC bus-based (top) and DC bus-based (bottom) EV charging station configuration. (*After:* [1].)

converter interfaces with a low-frequency transformer on the grid side and the DC-DC converter, which provides the DC power to charge the EV [1]. The first stage of an off-board EV charging system is the AC-DC converter, which performs two main functions: (1) acts as an interface between the power grid and the DC link and (2) controls reactive power consumption and handles grid-side current harmonics that are reduced through filters that act to isolate the converter from the grid. AC-DC converters can be unidirectional or bidirectional and can be single-stage or multistage power converters. As suggested by their names, multistage power converters combined multiple (two or more) power converters to achieve relatively higher power levels and a wider operating voltage range. Most off-board charging stations in the field today are designed with two-level or three-level voltage source converters that include buck/boost or multilevel converters. A number of different AC-DC converter designs are suitable for use in the EV charging station. These include:

- *Bridgeless boost converters:* They are composed of AC to DC diode rectifiers followed by a boost converter. These converters do not utilize a rectifier bridge used by other converter topologies to overcome the high conduction losses resulting in improved efficiency. The boost converter in these designs replaces the bridge rectifier with switches and also employs an AC side boost inductor [21].
- *Vienna rectifiers:* These operate in a continuous conduction mode that uses multilevel switching, which results in reduced voltage stressed on the power devices. A Vienna rectifier consists of three converters for the boost stage on the input side, three power bridges for the three phases, and two series output split capacitors on the DC link. The power flow for this converter is unidirectional, and each power bridge comprises two fast rectifier diodes and two reverse series connected switches.
- *Neutral point clamped converter*: These converters are characterized by the use of clamping diodes to ensure the proper sharing of voltages between the power switches. The switching losses are reduced by blocking all switches to half of the DC-link voltages.

The DC-DC converter is directly connected to the EV battery. The DC-DC converter converts the intermediate DC voltage into a desired regulated DC voltage level to charge the EV battery. DC-DC converter designs can be isolated or nonisolated. A number of different DC-

DC converter designs are suitable for use in the EV charging station. These include:

- *Interleaved buck converters:* Interleaving is a technique in power converter design for switching multiple power converter stages in parallel. Interleaving involves connecting multiple converter stages to the same voltage source and sharing the same power bus and one or more outputs. Interleaving can be symmetric where the switching of each stage of the power converter is separated by some phase difference or asymmetric where the PWM signals used to drive each stage of the power converter do not have a phase or frequency relationship [22].
- *Full-bridge LLC resonant converters:* The full-bridge LLC resonant converter combines two conversion topologies: the full-bridge topology and the LLC resonant converter topology. The converter uses a switching bridge that generates a square wave, which is then used to excite the LLC resonant tank. The LLC resonant tank outputs a resonant sinusoidal current that gets scaled and rectified by the transformer and rectifier circuit [23].
- *PSFB converters*: A PSFB converter consists of four switches that form a full bridge on the primary side of an isolation transformer and diode rectifiers or MOSFET switches for synchronous rectification on the secondary side. This topology allows all the switching devices to switch, with ZVS resulting in lower switching losses and an efficient converter [20].

## 4.3 RISKS AND LIMITATIONS

The high operating voltages and relatively high operating currents of charging stations introduce the risk of electric shock and burn hazards with both OBCs and off-board chargers. Understanding these hazards and identifying mechanisms to respond to failure modes become important for EV chargers that can operate at several hundred volts and expose a user to this voltage in the event of a failure or that can lead to elevated temperatures on the charge cables and connectors due to the high operating currents.

## 4.3.1 Electric Shock Hazards

### *4.3.1.1 Understanding DC Voltage-Related Hazards*

"For a given current path through the human body, the danger to persons depends mainly on the magnitude and duration of current flow " [24]. However, when designing measures of protection against electrical shock, it is not practical to use time/current values to determine the level of protection needed. The criterion typically used for evaluating the risk of electrical shock is Touch Voltage, defined as the "product of the current through the body called touch current and the body impedance" [24]. The relationship between the current and voltage is not typically linear because the impedance of the human body varies with touch voltage and the different parts of the body that the current flows through. In addition, the value of the body impedance depends on factors such as duration of current flow, magnitude of touch voltage, frequency, degree of moisture on the skin, surface area of contact, pressure exerted, and temperature.

When discussing electrical shock hazards, standards typically divide the hazards into the following thresholds:

- *Threshold of perception:* This is the minimum value of current that causes any sensation for the person through which it is flowing.
- *Threshold of reaction:* This is the minimum value of current that causes involuntary muscular contraction.
- *Threshold of let-go*[10]*:* This is the maximum value of touch current at which a person holding electrodes can let go of the electrodes.
- *Threshold of ventricular fibrillation:* This is the minimum value of touch current through the body, which causes ventricular fibrillation.

---

10. IEC 60479-1 defined the threshold of let-go as the maximum value of touch current at which a person holding electrodes can let go from the electrodes. According to IEC 60479-1, there is no definable threshold of let-go for DC voltages. This is because only making and breaking of current lead to painful and cramp-like contractions of the muscles.

#### 4.3.1.2 *Time/Current Zones and Touch Voltage Thresholds for Ventricular Fibrillation*

Figure 22 in IEC 60479-1 includes a chart depicting the time/current zone of effects of DC currents in persons for a longitudinal upward (from hand to feet) current path.[11] Zone DC-3 in this figure, as defined in Table 4.2 [24], is divided by curves a, b, and c, which are defined as follows:

- *a:* This curve is the level beyond which a startle reaction of current becomes possible.
- *b:* This curve shows the lower boundary of the current level beyond which more serious and undesirable physiological effects begin to occur.
- *c:* This curve shows the level beyond which the likelihood of ventricular fibrillation begins to become a concern.

Table 4.2 describes the physiological effects associated with the zones defined in Figure 22 of IEC 60479-1.

**Table 4.2**
Physiological Effects Associated with the Zones in Figure 22 in IEC 60479-1

| Zone | Current Magnitude | Physiological Effects |
|---|---|---|
| DC-1 | Up to 2 mA | Slight pricking sensation possible when making, breaking, or rapidly altering current flow |
| DC-2 | From 2 mA to curve b | Involuntary muscular contractions likely, especially when making, breaking, or rapidly altering current flow but usually no harmful electrical physiological effects |
| DC-3 | Curve b and above | Strong involuntary muscular reactions and reversible disturbances of formation and conduction of impulses in the heart may occur, increasing with current magnitude and time; usually no organic damage expected |
| DC-4 | Above curve $c_1$ | Pathophysiological effects may occur such as cardiac arrest, breathing arrest, and burns or other cellular damage; probability of ventricular fibrillation increases with current magnitude and time |
| | $c_1$–$c_2$ | Probability of ventricular fibrillation increases by up to 5% |
| | $c_2$–$c_3$ | Probability of ventricular fibrillation increases by approximately 50% |
| | Beyond $c_3$ | Probability of ventricular fibrillation increases by more than 50% |

*From:* [24].

11. The current magnitude of the different thresholds in Figure 22 of IEC 60479-1 will depend upon the path of current flow through the body. The figure provides an example of thresholds for current flow from hand to feet.

The time/current zones shown in Figure 22 of IEC 60479-1 are for a longitudinal current path from hand to feet. The time/current zones change if the current takes a different part through the body. For example, Figures 1 and 2 in IEC/TR 60479-5 show the physiological thresholds for both AC and DC through two different parts of the human body (i.e., flowing from hand-to-hand and flowing from both hands to both feet).

Table 4.3 [25, 35] provides the perception/reaction, muscular reactions, and ventricular fibrillation threshold DC current values for different current paths through the body. As can be seen in Table 4.3, the threshold current magnitudes for ventricular fibrillation depend strongly on the path that the current takes through the body, and this must be factored into when performing a risk analysis of any charger and evaluating the hazards in the event of a failure.

#### *4.3.1.3 Relevant Standards*

Several standards provide electric shock threshold voltages and detail the requirements for protection boundaries around exposed energized conductors. This section summarizes some of the standards. The following discussion assumes the worst-case scenario (i.e., exposed movable conductors). In addition, this discussion is limited to DC voltages up to 500V.

**Table 4.3**
Current Threshold Values for the Three Types of Thresholds for Current Flow for 10 Seconds

| Type of Threshold | Current Path | Current Magnitude (mA) |
|---|---|---|
| Perception/Reaction | Hand-to-hand | 2 |
| | Both hands to feet | 2 |
| | One hand to feet | 2 |
| Strong muscular reactions | Hand-to-hand | 25 |
| | Both hands to feet | 25 |
| | One hand to feet | 25 |
| Ventricular fibrillation | Hand-to-hand | 350 |
| | Feet-to-both-hands | 140 |
| | Seat-to-one-hand | 200 |

*Source:* [25, 35].

#### *4.3.1.4 NFPA 70E: Standard for Electrical Safety in Workplace (2018)*

Article 130.2(A)(3) of NFPA 70E states the following:

> Equipment operating at less than 50V: Energized electrical conductors and circuit parts that operate at less than 50V shall not be required to be de-energized where the capacity of the source and any overcurrent protection between the energy source and the worker are considered and that it is determined that there will be no increased exposure to electrical burns or to explosion due to electric arcs.

NFPA 70E further defines two shock protection boundaries: limited approach and restricted approach. These are defined as follows:

- *Limited approach boundary:* An approach limit at a distance from an exposed energized electrical conductor or circuit part within which a shock hazard exists. When one or more unqualified persons are working at or close to the limited approach boundary, the designated person in charge of the workspace where the electrical hazard exists shall advise the unqualified person(s) of the electrical hazard and warn him or her to stay outside of the limited approach boundary.
- *Restricted approach boundary:* An approach limit at a distance from an exposed energized electrical conductor or circuit path within which there is an increased likelihood of electrical shock, due to electric arc-over combined with inadvertent movement. No qualified person shall approach or take any conductive object closer to exposed energized electrical conductors or circuit parts than the restricted approach boundary unless one of the following applies:
  - The qualified person is insulated or guarded from energized electrical conductors.
  - The energized electrical conductors or circuit parts are insulated from the qualified person.

#### *4.3.1.5 UL Standards*

Several UL and IEC standards provide voltage magnitudes that are considered to be hazardous live. These UL standards are predominantly geared towards the safety of electrical equipment and hence address shock hazards associated with exposed energized parts of electrical equipment. For example, UL 61010-1 [25] defined hazardous live as follows: "Hazardous live: capable of rendering an electric shock or electric burn in normal condition or single fault condition."

The standard states the following about hazardous live voltage levels: Voltages above the following levels in normal condition are deemed to be hazardous live. The AC voltage levels are 30V rms and 42.4V peak and the dc level is 60V. For equipment intended to be used in WET LOCATIONS, the AC voltage levels are 16V rms and 22.6V peak or 35 Vdc.

UL 60950-1 [26] states the following: "Steady state voltages up to 42.4-V peak, or 60 Vdc are not generally considered as hazardous under dry conditions for an area of contact equivalent to a human hand."

#### *4.3.1.6 IEC Standards*

Similar to the UL standards, IEC 61010-1 [27] defined a hazardous live condition with the following voltage levels (emphasis added): "The ac voltage levels are 33-V rms 46.7-V peak and the dc level is 70V. For equipment intended to be used in WET LOCATIONS, the ac voltage levels are 16V rms and 22.6 V peak or 35 Vdc."

IEC 60950-1 [28] stated the following: "Steady state voltages up to 42.4-V peak, or 60 Vdc, are not generally regarded as hazardous under dry conditions for an area of contact equivalent to a human hand."

As discussed in the previous sections, the high operating voltages of EV chargers introduce the risk of an electric shock hazard if not properly managed. Several EV chargers/accessories have been recalled due to the risk of electric shock hazards over the past few years. These recalls have been due to several reasons ranging from incorrect software preventing the charger from terminating charge in the event of a lost ground connection, isolation components that act as single points of failure, and physically damaged charge connectors.

### 4.3.2 Burn Hazards

The relatively high operating voltages and currents passing through the charge cables and the charge connectors can also introduce the risk of a burn hazard if the temperature management system is not properly designed or due to a deterioration or failure of insulating schemes.

#### 4.3.2.1 *The Science of Burn Hazards*

When a hot object is brought in contact with a cold object, heat flows from the hot object to the cold object [29]. The rate of this heat flow depends on a number of factors, including the difference in temperature between the two objects and the thermal resistance of each object. Theoretically, the heat flow will continue until the two objects are at thermal equilibrium (i.e., at the same temperature). When a user touches a hot surface, heat flows from the device's surface to the skin, causing the skin temperature to rise. The skin temperature does not rise immediately to the temperature of the hot spot. The rate of rise of skin temperature in this scenario depends on factors that include the temperature of the hot spot, the rate of heat flow from the hot spot to the skin, the heat stored in the hot spot, the time of contact with the hot spot, the contact pressure, and the rate of heat carried away by the blood in the body. As an example, a metal at an elevated temperature feels hotter when touched than does a plastic at the same temperature. This is because metals have comparatively lower thermal resistance than plastics, allowing a higher rate of heat transfer to the skin.

#### 4.3.2.2 *Skin Temperature*

When a user touches a hot spot, the outer layer of the skin (epidermis) makes contact with the hot spot, which causes the skin temperature to rise. As the skin temperature increases, the skin suffers first-degree burns, followed by second-degree burns and finally third-degree burns. Table 4.4 [30] lists approximate skin temperatures for the different degrees of skin burn.

#### 4.3.2.3 *Relevant Standards*

Several standards provide guidance on the maximum safe temperature for various materials. As an example, the European standard BS EN 13202:2000 [30] has provided charts relating surface temperature and contact time to burn threshold for a variety of materials (Table 4.5 [30]).

Another industry standard for information technology equipment, UL 60950-1 [31], provides touch-temperature limits for parts in operator access areas (Table 4.6 [31]).

**Table 4.4**
Various Degrees of Skin Burn

| Degree of Burn | Skin Temperature (°C) | Effect |
|---|---|---|
| First | 44–55 | Damage to outer layer of skin; burn heals on its own |
| Second | 55–60 | First layer of skin burns through, and damage to the second layer of skin occurs; however, the burn does not pass through to underlying tissues |
| Third | 60 | Involves damage to all layers of the skin |

*From:* [30].

**Table 4.5**
Burn Threshold Spreads for a Contact Period of 0.5 Seconds

| Material | Burn Threshold Spread for a Contact Period of 0.5 Second (°C) |
|---|---|
| Bare (uncoated) metal | 67–73 |
| Ceramics, glass, and stone | 84–90 |
| Plastics | 91–99 |
| Wood | 128–155 |

*From:* [30].

**Table 4.6**
Maximum Temperature Specified in UL 60950-1

| Parts in Operator Access Areas | Maximum Temperature (°C) | | |
|---|---|---|---|
| | Metal | Glass, Porcelain, and Vitreous Material | Plastic and Rubber |
| Handles, knobs, grips, held or touched for short periods only | 60 | 70 | 85 |
| Handles, knobs, grips, continuously held in normal use | 55 | 65 | 75 |
| External surfaces of equipment that may be touched | 70 | 80 | 95 |
| Parts inside the equipment that may be touched | 70 | 80 | 95 |

*From:* [31].

Although the various industry standards provide useful guidelines for determining maximum acceptable surface temperatures, the diversity of materials and configurations in EV chargers may require additional investigation to determine the correct threshold for surface temperatures and characterize the potential for burn hazards.

### 4.3.3 Fire Hazards

The high operating voltages and currents associated with EV chargers make them a potential fire hazard in the event of a failure that is not mitigated by the design of the chargers. Several EV chargers/accessories have been recalled due to the risk of fire hazards over the past few years for a variety of reasons ranging from improperly spliced connectors [32] to faulty charger wiring.

### 4.3.4 Cybersecurity-Related Issues

EV charging stations not only have high-power delivery capability but must also communicate to authorize charging, have a well-defined sequence for the charging process, and manage load. With the many involved stakeholders including software and equipment vendors, the cybersecurity-related risks due to poorly implemented EVSE can have significant financial impacts. Some researchers have also identified specific risks associated with different components of the EV charging ecosystem. These include [33]:

- EV connectors serving as targets of attackers due to the communication protocols allowing the attackers to introduce malware or change the charge settings.
- Hacking of user terminals allowing potential disruptions for EV owners, manipulations of pricing mechanisms, or causing unauthorized modifications to the equipment that may introduce safety risks.
- Security risks associated with internet-connected EV chargers.

A July 2022 Sandia National Laboratories report [34] identified the following risks associated with current EV chargers:

- Inadequate physical protection against unauthorized access;
- Posting of login and provisioning credentials inside the charger;

- Lack of encrypted hard drives inside the chargers;
- Lack of verification of firmware prior to its deployment on the charger;
- Enabling of debug and unused services.

The report listed a number of recommendations for EV charger security to address these risks and recommended the development of "effective perimeter defenses to protect EV charger assets" including firewalls, access control lists, and data-in-flight requirements (e.g., encryption, node authentication). Cybersecurity-related issues for EV charges are still evolving and a subject of significant research.

## References

[1] Acharige, S. S. G., et al., "Review of Electric Vehicle Charging Technologies, Standards, Architectures, and Converter Configurations," *IEEE Access*, Vol. 11, April 14, 2023.

[2] https://www.nio.com/nio-power.

[3] https://www.transportation.gov/rural/ev/toolkit/ev-basics/charging-speeds.

[4] https://www.energy.gov/eere/articles/pump-charge-extreme-fast-charging.

[5] https://www.power-sonic.com/blog/guide-to-level-2-ev-charging.

[6] https://www.power-sonic.com/blog/ev-charging-connector-types.

[7] SAE Electric Vehicle and Plug in Hybrid Electric Vehicle Conductive Charge Coupler J1772_201710.

[8] Electric Transportation Engineering Corporation, "Electric Vehicle Charging Infrastructure Deployment Guidelines for the Oregon I-5 Metro Areas of Portland, Salem, Corvallis and Eugene," 2010.

[9] https://www.technologyreview.com/2023/05/17/1073265/how-5-minute-battery-swaps-could-get-more-evs-on-the-road/.

[10] https://ample.com/.

[11] https://www.littelfuse.com/industries/electric-vehicle-charger.aspx.

[12] Elghanam, E., M. Hassan, and A. Osman, "Design of a High Power, LCC-Compensated, Dynamic, Wireless Electric Vehicle Charging System with Improved Misalignment Tolerance," *Energies*, Vol. 14, 2021, p. 885.

[13] Sagar, A., et al., "A Comprehensive Review of the Recent Development of Wireless Power Transfer Technologies for Electric Vehicle Charging Systems," *IEEE Access*, Vol. 11, 2023, pp. 83703–83751.

[14] Mohamed, N., et al., "A Comprehensive Analysis of Wireless Charging Systems for Electric Vehicles," *IEEE Access*, Vol. 10, 2022, pp. 43865–43881.

[15] Regensburger, B., et al., "High-Performance Capacitive Wireless Power Transfer System for Electric Vehicle Charging with Enhanced Coupling

Plate Design," *2018 IEEE Energy Conversion Congress and Exposition (ECCE)*, Portland, OR, 2018, pp. 2472–2477.

[16] Mou, X., et al., "Survey on Magnetic Resonant Coupling Wireless Power Transfer Technology for Electric Vehicle Charging," *IET Power Electronics*, Vol. 12, 2019, pp. 3005–3020.

[17] Joseph, P. K., E. Devaraj, and A. Gopal, "Overview of Wireless Charging and Vehicle-to-Grid Integration of Electric Vehicles Using Renewable Energy for Sustainable Transportation," *IET Power Electronics*, Vol. 12, 2019, pp. 627–638.

[18] Tao, H., et al., "Onboard Charging DC/DC Converter of Electric Vehicle Based on Synchronous Rectification & Characteristic Analysis," *Hindawi Journal of Advanced Transportation*, Volume 2019, Article ID 2613893

[19] https://chargedevs.com/features/a-closer-look-at-power-factor-correction/.

[20] Texas Instruments, Phase-Shifted Full Bridge DC/DC Power Converter Design Guide, May 2014.

[21] Jaiswal, D., et al., Analysis of Bridgeless PFS Boost Converter," *International Journal of Engineering Research and Technology*, Vol. 1, No. 5, July 2012.

[22] Texas Instruments Incorporated, "Optimal Design for an Interleaved Synchronous Buck Converter Under High Slew-Rate, Load-Current Transient Conditions," SLYT139, Feburary 2001.

[23] Infineon Technologies North America (IFNA) Corp., "Resonant LLC Converter: Operation and Design," Application Note AN 2012-09, V1.0, September 2012.

[24] IEC TS 60479-1 (2005-07), "Effects of Current on Human Beings and Livestock – Part 1: General Aspects," July 2005.

[25] UL 61010-1, "Electrical Equipment for Measurement, Control, and Laboratory Use - Part 1: General Requirements," 2012.

[26] UL 60950-1, "Information Technology Equipment – Safety – Part 1: General Requirements," 2007, p. 17.

[27] IEC 61010-1 (2010), "Safety Requirements for Electrical Equipment for Measurement, Control and Laboratory Use – Part 1: General Requirements," 2010, p. 41.

[28] IEC 60950-1 (2001), "Information Technology Equipment – Safety – Part 1: General Requirements," 2001, p. 35.

[29] Nute, R., "Principles of Thermally-Caused Injury," *IEEE PSES The Product Safety Engineering Newsletter*, Vol. 3, No. 3, pp. 6–12.

[30] BS EN 13202:2000, "Ergonomics of the Thermal Environment - Temperatures of Touchable Hot Surfaces - Guidance for Establishing Surface Temperature Limit Values in Production Standards with the Aid of EN563," Annex B, Table B.1, 2000.

[31] UL 60950-1, Second Edition, "Information Technology Equipment – Safety – Part 1: General Requirements, Table 4C – Touch Temperature Limits," 2007, http://ulstandards.ul.com/standard/?id=60950-1_1.

[32] https://static.nhtsa.gov/odi/rcl/2023/RCLRPT-23E034-8939.PDF.

[33] Hamdare, S., et al., "Cybersecurity Risk Analysis of Electric Vehicle Charging Stations," *Sensors*, Vol. 23, 2023, p. 6716.

[34] Sandia National Laboratory, "Cybersecurity for Electric Vehicle Charging Infrastructure," SAND2022-9315, July 2022.

[35] IEC/TR 60479-5 Technical Report (2007-11), "Effects of Current on Human Beings and Livestock – Part 5: Touch Voltage Threshold Values for Physiological Effects," November 2007.

# 5

# MOTORS

## 5.1 INTRODUCTION

A variety of electric motors have characteristics that make them suitable candidates for use in EVs. The role of the motor in an EV application is to convert the electrical energy stored in the EV's traction battery into mechanical energy through rotational motion to move the EV forward. The motor can also play the role of a generator by capturing the energy from the vehicle and charging the traction battery as the vehicle slows down. This chapter will start with a discussion of the relevant electromagnetic principles for electric motors. The second part of the chapter will discuss the operation and characteristics of different motor types and how they can be used in an EV application.

## 5.2 ELECTROMAGNETIC WORKING PRINCIPLES

At a high level, the main working principle of all electric motors relies on the interaction between the stator and the rotor. The stator, which is a stationary component, supplies a magnetic field either using permanent magnets or electromagnets. The rotor, which is a rotating component and placed in between the magnets of the stator, contains windings made of an electrical conductor, usually copper. When a current flows through the windings of the rotor, an electromagnetic force is applied to the windings, which, in turn, creates a torque and allows the rotor to rotate. The force experienced by the rotor windings, which are current-carrying wires, is known as the Lorentz force and is defined by the cross-product in (5.1) when under a homogeneous magnetic field:

$$\vec{F} = I\vec{l} \times \vec{B} \tag{5.1}$$

where $I$ is the current in the wire, $\vec{l}$ is a vector with a magnitude equal to the length of the wire and direction is along the wire, and $\vec{B}$ is a vector with a magnitude equal to the magnetic field.

The Lorentz force varies along the length of the rotor winding due to the shape of the winding. As an example, Figure 5.1 shows a current carrying coil placed between two magnets. The Lorentz force experienced on one side of the current-carrying coil is equal in magnitude but opposite in direction to the force experienced in the opposite side of the winding. This results in a rotation of the winding and the creation of torque. The resultant torque is proportional to the current in the coil and, as can be seen from (5.1), the maximum force on the current-carrying coil is achieved when the direction of the current and the direction of the magnetic field are perpendicular to one another.

A homogenous magnetic field can be achieved by using a permanent magnet as shown in Figure 5.1 or using electromagnets embedded into the stator. An electromagnet is composed of conductor windings wrapped around a magnetic core. The application of current to the conductor windings of the electromagnet results in a magnetic field, which can be determined by Ampere's law, as shown in (5.2):

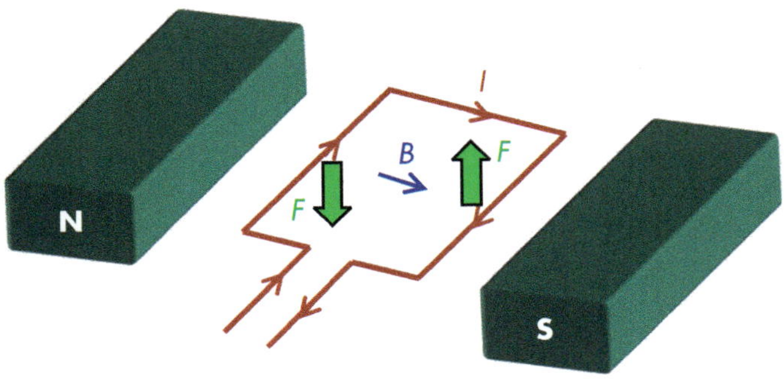


**Figure 5.1** Lorentz force applied to rotor winding.

$$\oint \vec{B} \bullet d\vec{l} = \mu_0 I_{enc} \tag{5.2}$$

where $I_{enc}$ is the current in the wire enclosed by an Amperian loop, $\vec{B}$ is a vector with a magnitude equal to the magnetic field, and $\mu_0$ is the permeability of free space also known as the magnetic constant, which quantifies how well a magnetic field permeates through a vacuum.

The loop on the integral sign denotes Ampere's law to be a line integral of the magnetic field over a closed loop. Another equation used to determine the magnetic field due to an applied current is the Biot-Savart Law as shown in (5.3). This law is a more general interpretation of Ampere's law and applies in situations where the magnetic field is not constant along an Amperian loop.

$$d\vec{B} = \frac{\mu_0 I}{4\pi} \frac{d\vec{l} \times \hat{r}}{r^2} \tag{5.3}$$

where $d\vec{B}$ is the differential magnetic field due to the current $I$ flowing in the short segment of wire $d\vec{l}$ and $\hat{r}$ is the unit vector pointing from the differential length to the test point.

Another configuration by which rotation of the rotor can be achieved is through electromagnetic induction, which involves inducing the needed current in the rotor using field coils in the stator. Induction will be discussed in detail in the context of AC induction motors in Section 5.5. In this configuration, the stator is made up of a few electromagnets similar to in a DC motor. When AC power flows through the stator electromagnets, an associated alternating magnetic field is created as described by Ampere's law. Due to this alternating magnetic field, an induced voltage, and, ultimately, a current, known as an eddy current, is generated on the rotor windings. This induced voltage, or EMF, follows Faraday's law as shown in (5.4):

$$\begin{aligned} \Phi_B &= \vec{B}(t) \bullet d\vec{A} \\ \varepsilon &= -\frac{d\Phi_B}{dt} \end{aligned} \tag{5.4}$$

where $\Phi_B$ is the magnetic flux and $\varepsilon$ is the induced voltage or EMF.

The negative sign within Faraday's law is defined by Lenz's law and simply describes that the induced EMF and magnetic flux have opposing signs. Additionally, the induced voltage and current within the rotor windings will, in turn, also create a magnetic field as defined by Ampere's law, which will oppose the direction of the stator field and thus create a back EMF. Well-known American physicist David Griffiths described the formation of this back EMF opposing the originally formed EMF as: "Nature abhors a change in magnetic flux" [1].

Motor technical specifications typically include a parameter called back EMF constant that can be used to estimate the back EMF for a given motor speed. The potential difference across a motor winding can be calculated by subtracting the back EMF value from the motor's supply voltage. The motors are designed with a back EMF constant in such a way that when the motor is running at the rated speed, the potential difference between the back EMF and the supply voltage will be sufficient for the motor to draw the rated current and deliver the rated torque. If the motor is driven beyond the rated speed, the back EMF may increase substantially, thus decreasing the potential difference across the motor winding and reducing the motor's current drawn, which results in a drooping torque curve.

## 5.3 BRUSHED DC MOTORS

Brushed DC (BDC) motors are simple and can be designed to deliver milliwatts to megawatts of output which makes them desirable for a number of applications. The main components of BDC motors include [2]:

- A *stator* that generates a stationary magnetic field either by using a permanent magnet or a configuration of electromagnets.
- A *rotor* that is made up of one or more windings that produce a magnetic field when energized. The rotor turns due to the interaction between the magnetic poles of the rotor and opposing magnetic poles of the stator. The windings of the rotor are constantly energized in a sequence that prevents the magnetic poles generated by the rotor from overrunning the generator poles in the stator. This process of constantly energizing the rotor windings is called commutation.

- An *armature* that interacts with the magnetic field generated by the stator and creates the mechanical torque that causes the motor shaft to rotate.
- A *commutator* that reverses the direction of current flowing in the windings to ensure that the produced magnetic field is such that the motor shaft turns continuously.
- *Brushes,* which are typically made up of carbon or graphite material that are in electrical contact with the commutator. As the motor turns, the brushes slide over different segments of the commutator. These segments are attached to different rotor windings and this attachment scheme results in the generation of a dynamic magnetic field inside the motor when a voltage is applied across the brushes.

### 5.3.1 Types of Brushed DC Motors

The three main types of BDC motors include:

- A *permanent magnet motor* where a magnet is used to generate the magnetic field in the stator to operate the motor. One of the more significant benefits of permanent magnet motors is reduced energy loss during operation as the motors generate less heat.
- A *shunt-wound type motor* that uses a parallel connection between the field windings and the armature windings. This parallel connection results in the same applied voltage applied to both the windings during operation. The motors maintain a constant speed over a wide range of loads that makes them suitable candidates for applications that require stable and predictable performance. Some limitations of these motors include erratic operational speeds, especially under no-load conditions and the potential for high-speed operation that can damage the motor in the event of a disruption in the field winding circuit.
- A *series-wound type motor* uses a series connection between the field windings and the armature windings. This results in the same current flow through the field windings and the armature during operation resulting in a strong magnetic field

allowing the generation of comparatively higher torque even at low speeds. This characteristic makes these motors specifically suited for applications that require significant power for short durations. However, the motor speed can be relatively unstable at lighter loads making them unsuitable for operation in applications that require a consistent speed under different load conditions.

BDC motors are often driven in applications using an H-bridge circuit. To understand the principles of operation of an H-bridge circuit, it is instructive to understand the various possible connection states for the motor terminals. A BDC motor essentially has two power supply terminals and is driven by applying a voltage to these terminals. The four possible combinations of connections to these two terminals include:

1. No voltage supply connected to the two terminals.
2. A positive voltage supply and a negative voltage supply connected to the two terminals.
3. A positive voltage supply and a negative voltage supply connected to the two terminals in the opposite polarity to number 2.
4. The two terminals are connected together.

The circuit to realize the four states can be created as shown in Figure 5.2. This circuit is called an H-bridge circuit due to the resemblance of the circuit to the capital letter H. The switches (denoted as SW in Figure 5.2) in most applications are replaced by MOSFETs or

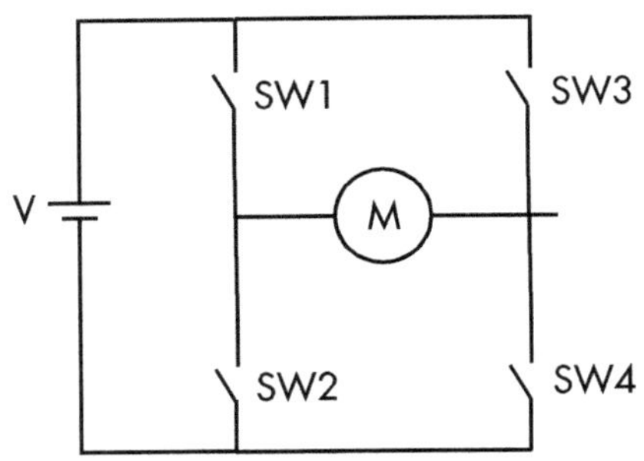


**Figure 5.2** H-bridge circuit used to drive a motor.

IGBTs to control the applied voltage and current flow through the motor.

While BDC motors are simple and have several advantages, the continuous contact between the brushes and the commutators gives rise to a few limitations that include [3]:

- A tendency to generate electrical and acoustic noise;
- The requirement for regular maintenance due to friction-induced wearing of the brushes, which results in a diminishing contact between the brushes and the commutator.

### 5.3.2 Torque Characteristics

The torque generated by a BDC motor is proportional to the motor current

$$T = K_t \times I_a \tag{5.5}$$

where $T$ is the motor torque, $K_t$ is the torque constant, and $I_a$ is the motor current.

This linearity makes it easy to calculate the operating parameters for a given motor under various conditions. In addition, for a constant input voltage, the speed of the DC motor falls linearly with an increase in the load (at maximum torque, which occurs at a zero rotation, the motor current is the maximum). In other words, the current increases as a function of the load applied to the output shaft. This allows a relatively straightforward determination of the current required for a specific motor to drive a desired load.

Taking into consideration the relationship between torque, motor speed, and input voltage, the following encompass the main electrical characteristics of a BDC motor:

- The speed of the motor:
  - Can be altered by changing the operative voltage or the power of the magnetic field. As an example, if a BDC motor rotates at 10,000 rpm at a constant supply voltage of 10V, the motor rotation will drop to 5,000 rpm at a 50% voltage duty cycle.
  - Is inversely proportional to the applied load.

- The linkage between the motor's magnetic field and its power supply can be used to control motor speed and torque and ensure a constant speed of operation.
- The number of coils surrounding the motor and the density of the coils determines the characteristics of the motor.

Figure 5.3 [4] shows the power and torque characteristics of a BDC motor suitable for use in a portable consumer electronic device. The power and torque characteristics of larger BDC motors suitable for use in EV applications are similar. As can be seen in Figure 5.3, the maximum torque occurs at lower speeds making the motor especially suitable for applications requiring high starting torque. However, the motor has a relatively lower efficiency at the lower speeds (and also relatively higher rotation speeds) resulting in relatively higher heat dissipation. The motor's power is the highest towards the middle of the motor's rated rotation speed where the efficiency of the motor is still relatively low. For the motor whose characteristics are shown in Figure 5.3, the maximum efficiency occurs at approximately 5,100

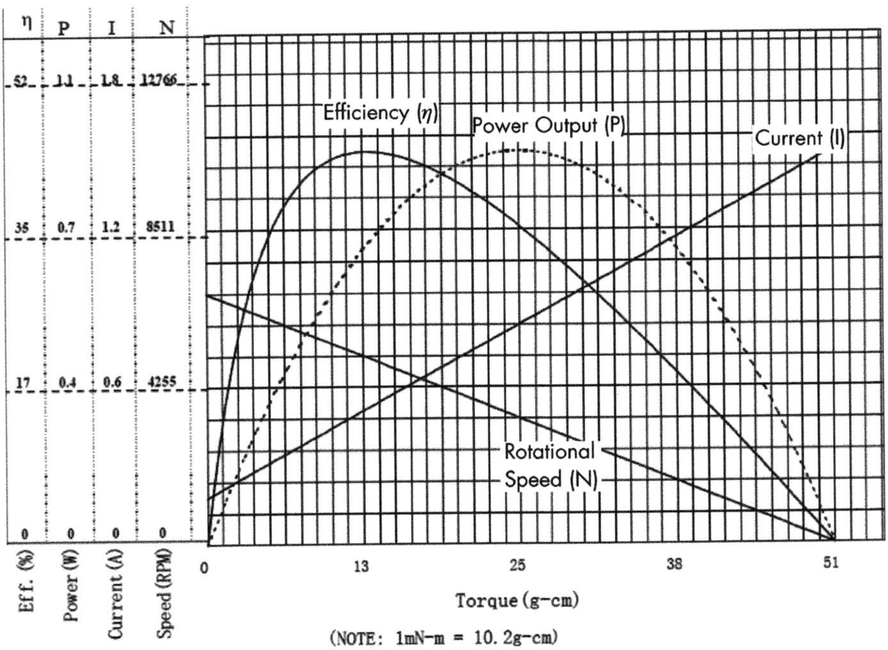


**Figure 5.3** Example BDC motor characteristics. (*From:* [4].)

rpm, but the torque and power of the motor at this rotation speed is relatively lower.

### 5.3.3 Motor Cogging

A common design for BDC motors uses a rotor that consists of a laminated iron core wound with wires to create the coils. Iron-core BDC motors can produce large amounts of torque but suffer from a phenomenon called cogging, which is caused by the attraction between the iron teeth of the rotor and the magnets of the stator. This attraction is present even when the motor is not powered. As the teeth of the armature pass over the edges of the stator magnets, the attraction between the two perturbs the rotation, which can result in a positioning error specially for motion control applications. Many techniques are used to reduce cogging such as skewing the teeth of the armature with respect to the edges of the magnets or using coreless designs.

## 5.4 BRUSHLESS MOTORS

As implicit in the name, brushless DC (BLDC) motors do not use brushes for commutation. The mechanical commutation in a BDC motor is replaced in a BLDC motor with switching achieved electronically using rotor position feedback sensors to determine when to switch the current. This electronic switching gives these motors several advantages including [5]:

- High-speed operation at full torque resulting in a relatively more powerful motor for a certain size as compared to a BDC motor where high speed operation at full torque shortens motor life due to the need for high energy to be switched by the brushes.
- The ability to mount the stator that contains the motor windings on a heat sink, which reduces temperature rise during operation and prolongs life.
- The relatively greater operating life when compared to BDC motor due to the absence of operating life limiting brushes.
- The ability to be used in applications that cannot operate in environments where brush particles generated during normal operation of a BDC motor are unacceptable.

- The absence of electrical and acoustic noise associated with the brushes of a BDC motor.

The main components of BLDC motors include [5]:

- A *stator*, which is typically a laminated steel core and carries the motor windings that are embedded and connected in one, two, or three phases. These phases are sequentially energized to create a rotating electromagnetic field during operation.
- A *rotor* that consists mainly of a shaft and a permanent magnet with alternating magnetic poles with the number of poles determined by the application in which the motor will be used. Most BLDC motors use an inner rotor design where the permanent magnet rotor is inside a wound stator, which increases the heat dissipation efficiency due to the winding position. In outer rotor designs, the external rotor houses the stator resulting in a relatively larger motor size. Figure 5.4 shows an example configuration of an inner rotor BLDC motor design.

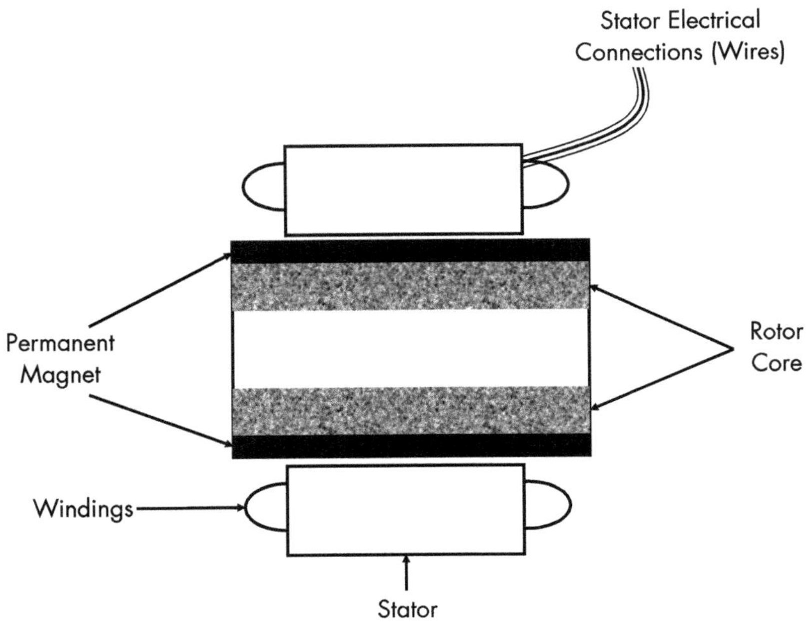


**Figure 5.4** Example configuration of an inner rotor design BLDC motor. (*After:* [5].)

- *Position sensors* that determine the position of the brushless rotor so that the excitation of the stator winding can be controlled to keep the motor rotating continuously. The following can be used to determine the position of the brushless rotor:
  - Photo electric sensors that use a set of phototransistors with light emitting diodes (LEDs) coupled across a shutter keyed to the rotor in a pattern that allows for the proper control of the phase excitation.
  - Electromagnetic sensors that use a soft iron target together with a set of wound coils where changes in the coil inductance are sensed and decoded to determine rotor position.
  - Magnetic sensors such as Hall-effect sensors, magneto-resistors or magneto-diodes that are incorporated directly in the stator assembly during manufacturing.
- *Control module* that receives information from the positions sensors and controls the excitation of the motor phases during operation through electronic switches (such as MOSFETs and IGBTs) using pulse width modulated (PWM) signals. The control module can be internal to the motor housing or placed outside the motor assembly.

Similar to BDC motors, BLDC motors also use H-bridge circuit designs to drive the motors. The H-bridge circuit designs can be identical to the H-bridge circuit designs in BDC motors for single phase BLDC motors. Three phase BLDC motors use half-bridge circuits with six switches in an arrangement where each phase of the motor is driven by a pair of high-side and low-side switches. Figure 5.5 shows a configuration where three half-bridges with common node coils are configured in a manner such that only two motor windings are excited at any one time.[1]

### 5.4.1 BLDC Motor Selection

The selection of a suitable BLDC motor for an application requires an understanding of the following three parameters [6]:

1. The three motor windings in this example are labeled U, V, and W.

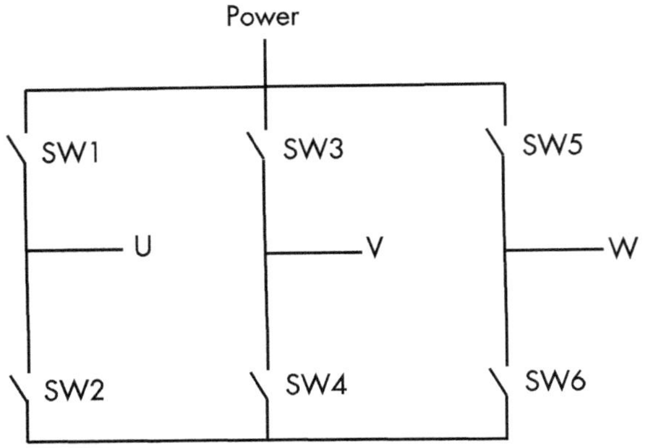


**Figure 5.5** Three half-bridges with common node coils.

- Peak torque ($T_P$): This can be calculated by summing the load torque, the torque due to inertia,[2] and the torque required to overcome friction. A safety margin (approximately 20%) is typically added to account for other losses in the system.
- RMS torque ($T_{RMS}$): This is the average continuous torque required for the application.
- Speed range: This is the required motor speed to drive the application.

Figure 5.6 [6] shows the typical torque-speed characteristics of a BLDC motor. The two parameters used to define the operation of the BLDC motor are peak torque ($T_P$) and the rated torque ($T_R$). During normal operation, the motor can be loaded up to the $T_R$. The motor's torque remains constant up to the rated speed of the motor. While the motor can be run at speeds exceeding the rated speed, the motor torque starts to drop as the motor speed increases. Some applications may demand more torque than the $T_R$ from the motor specifically under certain operating conditions such as when the motor starts from a standstill or as the motor accelerates to overcome the inertia of the load and the rotor itself. BLDC motors can deliver torque higher than the $T_R$ for brief periods of time to fulfill this torque demand as long

2. This is the torque required to accelerate the load from a standstill or from a lower to a higher speed.

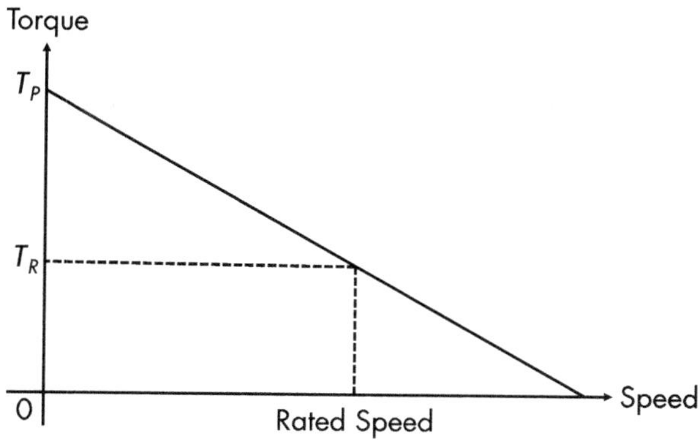


**Figure 5.6** Torque speed characteristics of a BLDC motor. (*After:* [6].)

as the speed-torque characteristics such as those shown in Figure 5.6 are met.

### 5.4.2 Comparing BDC and BLDC Motors

BLDC motors have several advantages over BDC motors due to their design. Table 5.1 [7] summarizes some of these advantages.

### 5.4.3 Brushless AC Motors

Unlike the alternative magnetic rotor poles that are energized by a DC in an BDC motor, brushless AC motors use three pole pairs in the motor with three passes of AC current, each shifted by 120° relative to each other. Similar to brushless DC motors, the shifting currents to each of the pole pairs cause the strongest point of the magnetic field to rotate. A smooth rotation of the magnetic field is achieved by applying a continuous AC current to the three pole pairs. Unlike BLDC motors where the rotor position is typically monitored by three Hall-effect sensors and where commutation is achieved in six steps or every 60 electrical degrees, brushless AC motors continuously monitor the rotor position via a single Hall-effect sensor or rotary encoder. This monitoring scheme allows the motors to operate without any torque ripple produced in BLDC motors due to each phase commutation (every 60°). Commutation in brushless AC motors requires more complex control algorithms.

**Table 5.1**
Comparing BLDC and BDC Motors

| | Advantages of BLDC Motors over BDC Motors |
|---|---|
| **Commutation** | Use of electronic switches results in comparatively higher motor life |
| **Efficiency** | A comparatively lower voltage drop on electronic switches as compared to the voltage drop on brushes results in superior efficiency |
| **Thermal Performance** | Superior thermal performance as the heat generated by the armature windings is connected to the outside case of the BLDC motor, which acts as a heat sink |
| **Maintenance** | No frequent requirement for maintenance due to the lack of brushes/ commutator in the motor |
| **Speed/Torque Characteristics** | Comparatively flatter due to the lack of brush friction, which can reduce useful torque |
| **Speed Range** | Comparatively higher due to the lack of mechanical limitations placed by the brushes or commutator |
| **Electric Noise** | Comparatively lower as no arcs from brushes to generate noise |
| **Dynamic Response** | Comparatively faster due to the use of permanent magnets, which result in lower rotor inertia |

*From:* [7].

## 5.5 AC INDUCTION MOTORS

As discussed previously, brushless AC motors work by using three-phase AC power to generate a continuously rotating magnetic field that pulls the magnetic poles of the rotor around in sync with the rotating field. The operating principle of an AC induction motor's stator is exactly the same as a brushless AC motor where it generates a continuously rotating magnetic field. The difference lies in the construction of the rotor, which, instead of being magnetically charged, is just a collection of loops of wire. As discussed previously, when a magnetic field rotates around a loop of wire, a current is induced in the wire and, due to the Lorentz Force law, the loop of wire experiences a force which for the AC induction motor is a torque in the direction of rotation of the magnetic field [8].

In contrast to BDC and BLDC motors, the stator in AC induction motors is composed of electromagnets or field coils that are powered by AC power and create an alternating magnetic field. In practice, just like with brushless AC motors, the stator field coils are each connected to a phase creating a three-phase system with each phase been 120° apart from one another. The use of a three-phase, AC power supply thus creates an alternating and rotating magnetic field. According to Faraday's law, this changing magnetic field will induce an EMF in the rotor windings. This induced EMF in the rotor creates a current and an accompanying varying magnetic field. Ultimately, it is the interaction between the induced magnetic field in the rotor and the magnetic field of the stator, which allows for rotation of the rotor.

The speed at which the stator magnetic field rotates is known as the motor's synchronous speed ($N_S$) and is defined as

$$N_S = \frac{120f}{P} \tag{5.6}$$

where $f$ is the power supply frequency and $P$ is the number of poles.

The rotor in an induction motor rotates in the same direction as the rotating magnetic field. However, the rotor's rotation speed is usually less than the synchronous speed. The difference between the rotor and synchronous speed is called the slip and is defined as shown in (5.7). Slip plays an essential role in the operation of induction motors and is required to produce torque. However, slip is also among the reasons why AC induction motors are not constant-speed motors and require the use of variable frequency drives to control the motor's speed. The amount of slip present in AC induction motors is approximately 2% to 3% [9, 10].

$$\%\,slip = \frac{N_s - N_b}{N_s} \times 100 \tag{5.7}$$

where $N_b$ is the base speed or rotor speed.

The advantages of AC induction motors, when compared to DC motors, include:

- Only one power supply is needed to drive the system. As discussed, in an AC induction motor, only the stator field coils need to be AC-powered to commence the induction process to the rotor. In contrast, in DC motors, often two power supplies are needed if electromagnets are used within the stator. One power supply is used to power the stator field coils and the other is used to power the rotor windings.
- Since the rotor windings do not require an external power supply, commutators and brushes are not needed to complete the

circuit. As discussed previously, the use of brushes introduces disadvantages such as reduced operating lifetime of the motor or the need for maintenance due to brush degradation during the lifetime of the motor.

- The reduction of components like brushes and commutators and the reduced need for maintenance also factors into cost-effectiveness, less noise, and longer operating lifetimes.

Conversely, one of the main disadvantages of AC induction motors is the lack of speed control. As shown in (5.6) and (5.7), the speed of the rotor is defined by the input power frequency, which ranges from 50 to 60 Hz. Advances in variable frequency drives have allowed this disadvantage to be addressed in certain applications.

### 5.5.1 Types of AC Induction Motors

For an EV application, three-phase AC induction motors are used and are typically classified in two categories [11, 12]:

- *Squirrel cage motor*: The components making up the rotor of this type of motor include a metal shaft, laminations, and a squirrel cage. In an EV application, the metal shaft is used to interface the rotor with the vehicle gears and ultimately move the vehicle forward. The laminations are thin metallic plates stacked together and are used to minimize the effect of eddy currents produced during the induction process. Lastly, the squirrel cage, named as it looks, is a short-circuited cylindrical metallic frame that sits around the rotor laminations and shaft, and it is this component that rotates when the motor is in use. Most three-phase induction motors used today within industrial and commercial applications are squirrel cage induction motors due to their simplicity, low cost, ease of maintenance, and ruggedness.
- *Wound rotor induction motor or slip-ring motor*: The rotor construction of a wound rotor induction motor differs from that of a squirrel cage rotor in that the rotor bars are not short-circuited. Instead, the rotor windings are connected to slip rings and brushes at each end allowing for the introduction of resistances or contactors to the system. The need for in-line resistances is due to the start-up characteristics of AC induction motors. Of-

ten, large starting currents (at times 5 to 10 times the full-load current) are needed to produce large enough torques which rotate the rotor to full speed. In the case of high-power electronics, these large currents can prove problematic due to the potential of hazards such as wire overheating or arcing. Adding in-line resistances allow for lower starting or in-rush currents. Additionally, since the motor resistance is proportional to the maximum torque produced by the motor, adding in-line resistances to the motor allows for tunability of the speed and torque characteristics of the motor ultimately allowing the motor to achieve maximum torque at lower motor speeds. Once the required characteristics are achieved for the load following start-up, the resistors can be removed, and the motor performs similarly to a standard induction motor. Although the wound rotor induction motor provides advantages for high-inertial load applications where fast motor speeds need to be achieved quickly, it also has its disadvantages. Due to the use of slip rings and brushes, increased maintenance is required for these motors as compared to the squirrel cage motors. Additionally, due to the addition of windings and resistors to the rotor assembly, these motors prove to have a relatively more complicated construction.

### 5.5.2 Speed-Torque Characteristics

As described in the previous section, AC induction motors require large torques to achieve full motor speeds during start-up. As shown in Figure 5.7 [10, 12], as the motor starts up, the pull-out torque needed for optimal motor operation is not usually achieved until the motor reaches 70% to 80% of its full motor speed. However, increasing the motor resistance, as with a wound rotor induction motor, does tune the speed-torque characteristics, effectively making it so the maximum torque needed for motor operation is achieved at a lower motor speed.

## 5.6 RELUCTANCE MOTORS

A reluctance motor is a motor that uses the principle of reluctance torque for operation. Reluctance torque is the torque induced in an

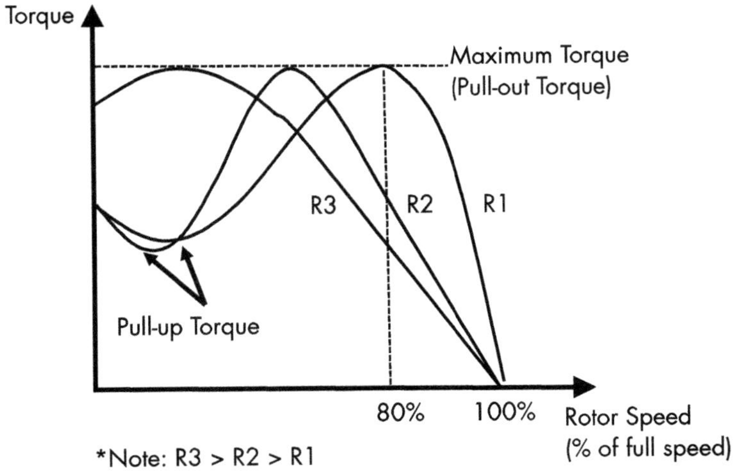


**Figure 5.7** AC induction motor speed-torque characteristics. (*After:* [10, 12].)

iron object in the presence of an external magnetic field. Reluctance motors polarize the amount of reluctance (the magnetic equivalent of electrical resistance) in their rotor poles to interact with the stator's rotating magnetic field. The two main types of reluctance motors use the principle of reluctance in their own way and include synchronous reluctance motors and switched reluctance motors. Reluctance motors have a relatively simpler construction due to the lack of slip rings, rotor field windings, brushes, or commutators, which improves efficiency, reliability, and maintenance costs. However, the relatively simpler design of these motors requires relatively complex electrical and control circuits for operation.

The two main components of a reluctance motor are the stationary stator and the rotor both of which are separated by a small air gap. The stationary stator of a reluctance motor consists of protruding salient pole-pairs[3] that are created by running current through a wire wound around these protrusions. The rotor is created using ferromagnetic metal with its own poles designed to follow the contours of the stator's magnetic field. When a rotor's pole lines up with a stator's pole, the rotor is in the minimum reluctance position (i.e., the magnetic resistance is the lowest at this point and the rotor is fully

3. The term "saliency" refers to the ability to project outwards. Salient pole machines have poles extending outwards.

aligned). When the stator's pole lines up with the notches, barriers, or slots in the rotor, the rotor is in the maximum reluctance position and fully misaligned. The rotor always moves to the position of least reluctance, hence resulting in the creation of a reluctance torque when the rotor is fully misaligned. This reluctance torque pulls the rotor to the nearest salient stator pole, resulting in rotation. The use of control systems and specific rotor geometry allows for the continuous rotation output from the motor [13].

There are two main designs of reluctance motors [13]:

- Synchronous reluctance motors that are designed to run at exact or synchronous speeds. These motors use a three-phase stator winding and a rotor that implements salient rotor poles and internal magnetic flux barriers that are in the form of notches or air gaps in the rotor. As the motor starts, it is pushed close to synchronous speeds and then locks in synchronism via the reluctance torque generated by the rotor flux barriers. The structure of the stator of synchronous reluctance motors is nearly identical to that of an induction motor. The rotor consists of a round laminated core from which magnetic flux barriers are stamped. The rotor is not suitable for high speeds as additional bars must be inserted into the flux barriers for stiffness during high-speed operation. However, these additional bars have a detrimental effect on the efficiency of the motor.
- Switched reluctance motors (SRM) are considered a type of stepper motor[4] that use a ferromagnetic core of permanent poles and notches that are acted upon by the electromagnetic poles of the stator. The stator poles are switched on and off by an electronic circuit to lead the rotor poles to each newly aligned position to allow the motor to rotate.

### 5.6.1 Switched Reluctance Motors

As discussed previously, SRMs can be considered a stepper motor type and like other reluctance motors, SRMs usually have no magnets in the stator or the rotor, thus resulting in a simple structure. How-

4. Note that SRMs have differences from stepper motors. These differences include that SRMs have comparatively bigger steps and requiring closed loop control system.

ever, some applications use a small amount of permanent magnet materials to improve the motor's torque. The number of the ratio of stator poles to rotor poles used in SRMs is limited with the most common being 4:6 and 6:8, each with its own possible multipliers [14].

SRMs can be manufactured with a range of structures depending upon the application. SRM classification is based on movement pattern, flux path and the type of excitation and include rotary and linear SRMs. Some limitations of SRMs include:

- SRMs have a relatively lower torque capacity and can also be noisy.
- Torque ripple, specifically at higher operating speeds, must be addressed in an application.
- The motor can generate harmonics at higher speeds that must be addressed through filtering capacitors.
- The motor has a comparatively large torque ripple and comparatively lower efficiency as compared to a synchronous reluctance motor.
- Larger phase currents resulting in control circuits that are comparatively more complex as compared to the control circuits for synchronous reluctance motors.

### 5.6.2 Specifications and Torque Characteristics

Some specifications that are common to all reluctance motors and that must be analyzed to determine the suitability of a selected reluctance motor for an application include:

- *Stator to rotor pole ratio:* Reluctance motors come in a variety of pole ratios. Some common pole ratios include 6:4, 4:2, 8:6, or 12:8. A 1:1 ratio works best for synchronous reluctance motors, as it allows for more precise synchronism.
- *Rated power and torque:* Just like for any motor, the rated power and torque of the motor provides an indication of the amount of output power and torque provided by the motor at a steady state.
- *Torque ripple:* Reluctance motors are susceptible to a torque ripple effect due to a periodic variation in the produced torque

that occurs when the stator and rotor poles are in the fully aligned positions. This torque ripple can result in noise and other oscillation in the system. For this reason, it is important to understand the rated torque ripple of the motor selected for the application.

- *Power factor:* Synchronous reluctance motors can have a comparatively lower power factor, which must be considered when evaluating the suitability of the motor for an application.

The performance of a SRM depends upon three control parameters: turn-on angle, turn-off angle, and the amplitude of the phase current. Figure 5.8 [15, 16] shows the SRM torque speed characteristics. As can be seen in Figure 5.8, the SRM can be made to operate in one of the following three modes by controlling these three parameters [16]:

- Constant torque mode;
- Constant power mode;
- Mode in which the motor torque is inversely proportional to the square of the motor speed.

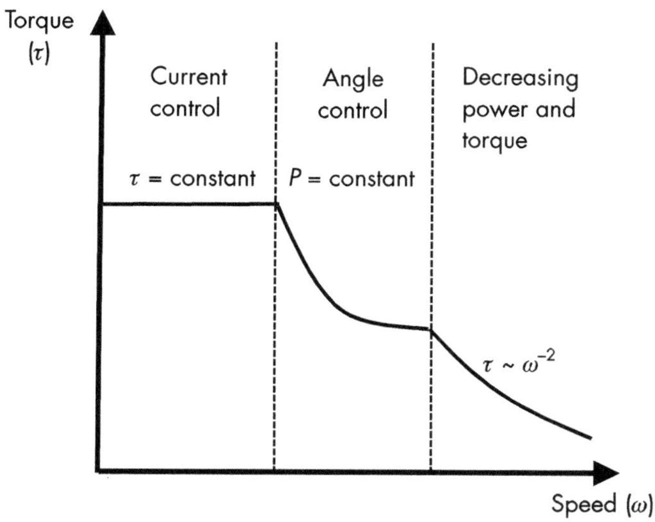


**Figure 5.8** SRM torque speed characteristic. (*After:* [15, 16].)

## 5.7 MOTOR EFFICIENCY

Efficiency, usually represented as a percentage, is defined as the ratio between the output power and input power of the motor. Succinctly, this value quantifies the amount of losses that the motor is expected to experience under normal operation or, in other words, how good the motor is at converting electrical energy into mechanical energy. Efficiency is directly proportional to the motor load and is calculated using (5.8) [17]:

$$Load = \frac{P_i \times \eta}{hp \times 0.7457} \tag{5.8}$$

Where *Load* is the output power as a percentage of the rated power, $P_i$ is the power in kW, $\eta$ is the efficiency in percentage, and *hp* is the rated horsepower.

The efficiency is usually calculated when the motor is manufactured, and this value is included in its nameplate for reference. However, as the motor ages, the nameplate efficiency might no longer match the true efficiency of the motor and thus new motor losses are introduced to the system over its lifetime. Motor losses fall in various categories including [18, 19]:

- *Conductive loss:* These include ohmic losses in the windings of the stator and rotor. When the motor operates at high temperatures, the resistance within these windings increases resulting in increased heat production as defined by Ohm's law $P = I^2R$. Conductive losses can also be frequency-dependent through what is known as the skin effect, which describes the tendency of AC current to distribute themselves along the surface of the conductor due to the presence of opposing eddy currents. This current distribution effectively reduces the cross-section of the conductor through which current can travel and thus increases the resistance of the wire again resulting in heat generation. The effective cross area of the wire decreases with increasing frequency. As such, effective cooling mechanisms must be implemented to maintain the motor at an optimal operating temperature.

- *Core losses:* The mechanisms for these losses are due to the presence of eddy currents, as described by Faraday's law, and magnetic field hysteresis, which occur due to constant magnetization and demagnetization of the iron core of certain motors.
- *Mechanical losses:* These losses can be broken up into two categories, which include frictional losses and windage losses. Frictional losses relate to increased friction, for example, on the bearings which results in heat generation. Windage losses relate to losses due to wind resistance on rotating components of the motor.
- *Other losses:* The efficiency of the motor might also decrease due to aging and physical deterioration of the unit or due to preventative maintenance performed in the unit such as rewinding of the motor.

Ultimately, as technology advances, the goal is to manufacture and implement high-efficiency motors into EVs. To achieve the latter, various considerations during the manufacturing and maintenance of these motors must be thought through. These include [20, 21]:

- The use of thinner laminates in the body of the rotor to further minimize the effect of eddy currents.
- Reduction in the frictional force imposed on rotating components through regular maintenance and greasing of the bearings.
- Optimized fan designs to address windage losses.
- Optimized cooling mechanisms to mitigate resistive losses.
- Improved electrical steel, and the use of high-quality materials in general.

## 5.8 MOTOR CHARACTERISTICS IN EVS

Figure 5.9 [22] shows the desired output characteristics of electric motor drives in EVs and the speed-torque characteristics of standard electric motors. EVs must operate in constant power for initial acceleration, which results in a desire to have a motor drive that has a high torque at low speeds. Once the EV has accelerated, the motor must

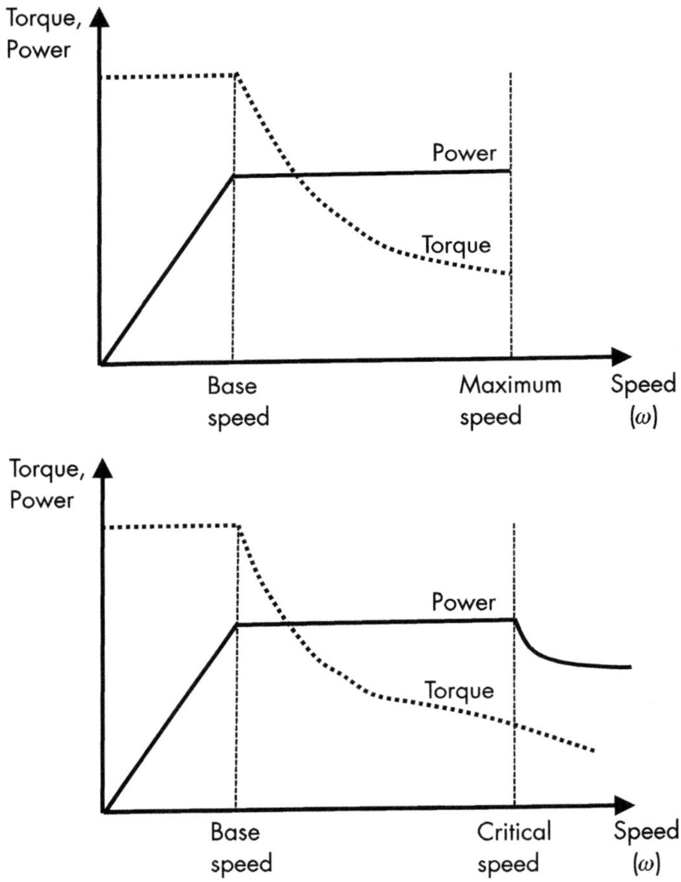


**Figure 5.9** Desired output characteristics for electric motor drives in EVs (top) and speed-torque characteristics of standard electric motors (bottom). (*After:* [22].)

provide a high power at a high speed for operation over a wide range of vehicle speed or while cruising at a higher speed. Motor drives in EVs are typically operated in the constant power region beyond a base speed up until the maximum speed of the motor. The motor torque is inversely proportional to the square of the speed, which also results in a reduction in the total motor power requirement.

Table 5.2 compares some features of the different motor types that are potential candidates for use in an EV application. Each motor type has its own advantages and limitations for EV applications. As an example, although BLDC motors have a high-efficiency and

**Table 5.2**
Comparing Motor Types for EV Applications

| Feature | Description | BLDC Motor | BDC Motor | AC Induction Motor |
|---|---|---|---|---|
| Speed vs. torque characteristics | Typically depicted in a chart containing information on the ability of a motor to provide torque at a certain rotational speed | Linear; for this reason, operation at all speeds with rated load possible | Approximately linear; however, the torque developed decreases at high speeds due to the friction of the brushes | Nonlinear with comparatively lower torque generated at low speeds |
| Commutation | Methodology used to drive motor current and switch the current between the motor phases | Electronic | Brushed | — |
| Rating | Operational limits of the motor | High | Low | Moderate/low |
| Efficiency | Ratio of the motor's mechanical and electrical power | High | Moderate | Low |
| Maintenance requirements | Dictated by the motor's brushes | Low | High | Moderate |
| Speed range | Rotational speed characteristics | High | Low | High |
| Electric noise | Refers to switch related electrical noise generated by the motor | Low | High | Low/moderate |
| Control | Circuitry used to control motor operation | Complex and expensive | Simple and inexpensive | Simple and inexpensive |
| Life | Expected motor lifetime under designed operating conditions | Long | Short | Moderate |
| Slip | Speed difference between motor synchronous and asynchronous operation | No | No | Yes |
| Starting current | Current required to start motor rotation | Rated | — | 5 to 10 times rated current |

*From:* [22].

high-speed range, they require sophisticated and complex driving circuitry, which increases the overall cost. BDC motors can be designed with relatively simple control circuits but are hampered by the need for commutator maintenance, which makes them less suitable for EV applications.

## References

[1] Griffiths, D., *Introduction to Electrodynamics*, Boston, MA: Pearson, 2013, p. 315.

[2] Reston Condit Mirochip Technology, Inc., "Microchip, Brushed DC Motor Fundamentals," AN 905, 2010, DS00905B, 2010.

[3] Hamilton, R. J., "DC Motor Brush Life," *IEEE Transactions on Industry Applications*, Vol. 36, No. 6, 2000, pp. 1682–1687.

[4] Shenzhen Shuanghuan QX Motor Co., Ltd., Specification Number QX-CM130335, Rev A0, 2018.

[5] General Dynamics, "Ordnance & Tactical Systems," *Brushless DC Motor Handbook*, 2020.

[6] Yedamale, P., "Microchip Technology Brushless DC (BLDC) Motor Fundamentals," AN885, 2003.

[7] Monolithic Power, "Brushless DC Motor Fundamentals," AN047 Rev 1.0, 2011.

[8] Enge, P., N. Enge, and S. Zoepf, *Electric Vehicle Engineering*, New York: McGraw-Hill, Chapter 4, 2021.

[9] https://www.nidec.com/en/technology/motor/basic/00027/.

[10] Parekh, P., "Microchip, AC Induction Motor Fundamentals," AN887, 2003, DS00887A.

[11] https://savree.com/en/encyclopedia/induction-electric-motor-squirrel-cage.

[12] https://www.thomasnet.com/articles/machinery-tools-supplies/what-is-a-wound-rotor-motor-and-how-does-it-work.

[13] Cavallo, C., ThomasNet, "All About Reluctance Motors - What They Are And How They Work," *Thomasnet,* August 19, 2020.

[14] Mahdi Bouiabady, M., A. D. Aliabad, and E. Amiri, "Switched Reluctance Motor Topologies: A Comprehensive Review," June 21, 2017, DOI: 10.5772/intechopen.69149.

[15] Liu, L., et al., "Parametric Modeling and Optimization of Switched Reluctance Motor for EV," *ACES Journal*, Vol. 37, 2023, pp. 948–958.

[16] Calasan, M., and V. Vujicic, "Characteristics of Switched Reluctance Generator Operating with Derishzadeh Converter," *International Conference of IcETRAN,* Zlatibor, Serbia, 2016

[17] US Department of Energy, "Determining Electric Motor Load and Efficiency," https://www.osti.gov/servlets/purl/1215875.

[18] Roshandel, E., et al., "Losses in Efficiency Maps of Electric Vehicles: An Overview," *Energies*, Vol. 14, 2021, p. 7805.

[19] Jenkins, J., "A Closer Look at the Losses in EV Motors," *Charged Electric Vehicles Magazine*, No. 41, January/February 2019.

[20] Kaya, D., F. Çanka Kılıç, and H. H. Öztürk, "Energy Efficiency in Electric Motors," in *Energy Management and Energy Efficiency in Industry: Green Energy and Technology*. New York: Springer, 2021.

[21] NEMA, "Making Electric Motors More Efficient," https://www.nema.org/docs/default-source/motor-and-generator-guides-and-resources-library/5-making-elec-motors-more-efficient-v2.pdf?sfvrsn=3343d49d_2.

[22] Nitin Subramonium A. K., et al., "Technology and Key Strategy of IE 4 Permanent Magnet Brushless DC Motor Drive for Electric Vehicle Application," *International Journal of Engineering Research and Application*, Vol. 7, No. 2, Part 4, February 2017, pp. 25–31.

# 6

# CONVERTERS

Connecting an EV's traction battery to its motor requires an intermediate circuit so that the battery power can be converted into a form that is suitable for driving the motor. This conversion may be a conversion to a different voltage, conversion from DC voltage to AC voltage, change in the frequency of the AC voltage, or all of the above. Similarly, during regenerative braking, the electricity generated by the EV's motor needs to be converted to a suitable form that can be used to charge the traction battery. Figure 6.1 shows an example of an EV powertrain architecture. Several variations of this architecture are found in hybrid vehicles and EV applications. In Figure 6.1, the bidirectional DC-DC converter boosts the battery voltage, so that a higher-speed motor can be used, which has a smaller size and higher efficiency. As an example, in a model year 2010 Toyota Prius, the nominal voltage of the hybrid battery is approximately 200V and the DC-DC converter boosts this voltage to 650V. The boosted voltage goes through a motor inverter that converts the DC voltage into an AC voltage appropriate to drive the AC motor.

A higher-level overview of some specific power converter architectures used in EV applications (DC-DC converters, AC-DC converters) has already been provided in Chapter 4. This chapter will discuss the operating fundamentals of the power converter architectures discussed in Chapter 4. This chapter will start with an overview of some of the components that are used as fundamental building blocks for power converters and will then provide a detailed discussion of how

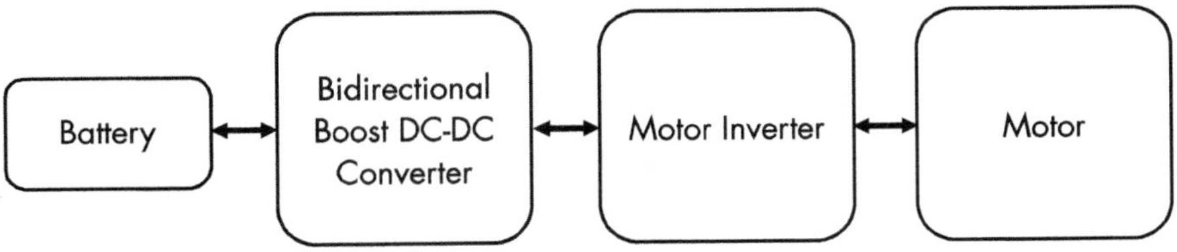


**Figure 6.1** Example EV powertrain architecture.

these components can be used to convert power from one form to another.

## 6.1 COMPONENTS

The main components that form the building blocks of a power converter include:

- Power metal-oxide-semiconductor field-effect transistors (MOSFET), insulated-gate bipolar transistors (IGBT), and/or thyristors, which act as switches that rapidly turn on and off following a specific duty cycle[1] to aid in the power conversion process.
- Power inductors that store energy when current is flowing through them and discharge this stored energy once the current ceases to flow.
- Capacitors that act as filters, help with voltage regulation, and provide transient protection.
- Power diodes that are used for rectification. Additionally, within DC-DC converters, diodes also serve the role of selecting the current path for the inductive currents to travel as the other power electronics are switched.

## 6.2 POWER MOSFETS

A MOSFET, as shown in Figure 6.2, is a semiconductor device that effectively acts as a switch and contains three terminals: source, drain,

1. Duty cycle is defined as a time ratio between when the circuit is on versus when the circuit is off.

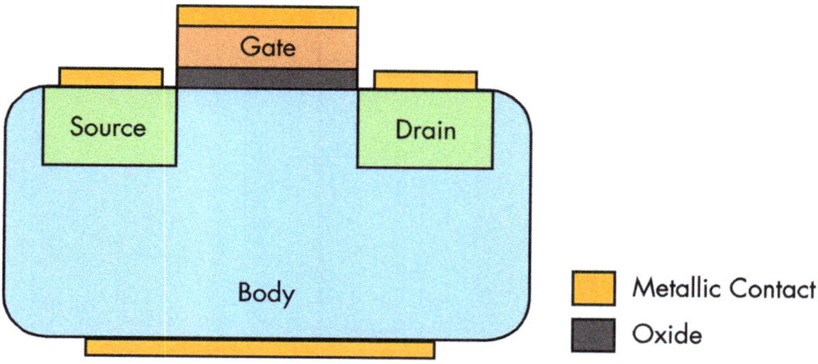


**Figure 6.2** Typical MOSFET construction.

and gate. The on/off state of the MOSFET is determined by the voltage that is applied at the gate. In technical terms, the conductivity within the MOSFET is modulated with the application of an electric field (i.e., voltage) at the gate of the device. This is known as the field effect. In the on state, current flows between the source and the drain of the MOSFET through a conductive channel that is created in the body of the MOSFET. Conversely, in the off state, this conductive channel is suppressed ceasing current flow.

The material that makes up the MOSFET is silicon (Si). In order to achieve enhanced conduction within Si such that it can be used as a switching device, impurities in the form of electron or hole donors must be added. These impurities are introduced by what are known as dopants. Depending on which carrier, electron or hole, is introduced into the Si through these dopants, the MOSFET will either be classified as N-channel, where electrons are the majority carrier, or P-channel, where holes are the majority carrier. N-channel MOSFETs are more commonly used due to their lower on-state resistance ($R_{DS}$) as compared to P-channel MOSFETs. N-channel MOSFETs also require a positive voltage to operate which is the natural polarity of ground-referenced (0V) switching devices.

N-channel and P-channel MOSFETs can be further categorized as depletion or enhancement mode MOSFETs (an example of an N-channel MOSFET in these two modes is shown in Figure 6.3). Depletion mode MOSFETs are normally closed switches, and the application of a gate voltage turns the MOSFET off. In contrast, enhancement mode MOSFETs are normally open switches, and the application of

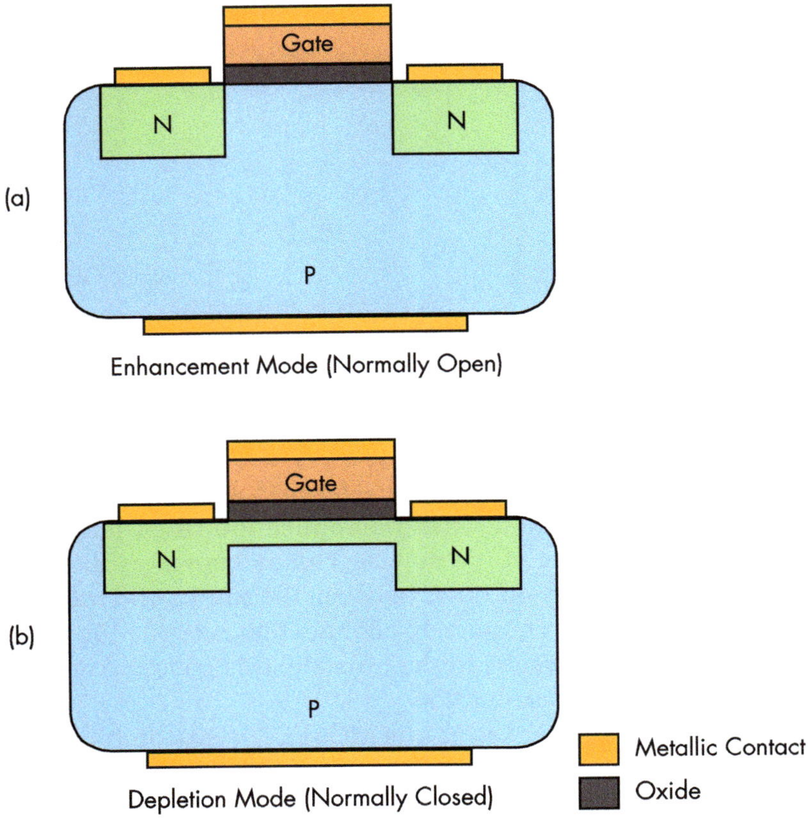


**Figure 6.3** N-channel MOSFET in (a) enhancement and (b) depletion modes.

a gate voltage turns the MOSFET on. Enhancement mode MOSFETs are more commonly used in power conversion applications due to their comparatively smaller geometry and lower energy dissipation [1]. Additionally, as the MOSFETs are normally open, the possibility for false triggering in switching circuits is lower because a voltage must actively be applied to close the switch.

Microscopically, the working principle within an enhancement N-channel MOSFET relies on the tunable conduction of electrons in the presence of an applied voltage. The addition of dopants creates localized regions of N-type Si near the source and drain terminals, while the remaining substrate is the opposite type as shown in Figure 6.3(a). When a voltage is applied to the gate, an electric field is formed that results in an accumulation of electrons near the gate that effectively creates an N-type channel that fully connects the source and drain and allows for the conduction of electrons. When the gate voltage is removed, the electrons repel one another and proceed to diffuse throughout the P-type substrate, effectively, suppressing this conductive transport channel.

Power MOSFETs are a subset of MOSFETs used for high-voltage and/or high-current applications such as in EV power conversion circuits. Power MOSFETs are designed to operate at high currents and voltages without exhibiting significant degradation or breakdown over numerous switching cycles. This high-current/voltage operation is achieved by using a construction that is different from that of a conventional lower power MOSFET as shown in Figure 6.4. Power MOSFETs are constructed in a vertical configuration with the source and drain located on the top and bottom of the MOSFET. Power MOSFETs also include an additional lightly doped N-type region between the source and drain that is called a drift region. The addition of this lightly doped region leads to the creation of a P+/N– junction and it is the interaction within this junction that enables a high-voltage operation. When the MOSFET is off, the P+/N– junction acts as a reverse-biased diode and when a voltage is applied a depletion region (no carriers present) is created which is contained within the drift region. The drift region dimensions determine the breakdown voltage of a

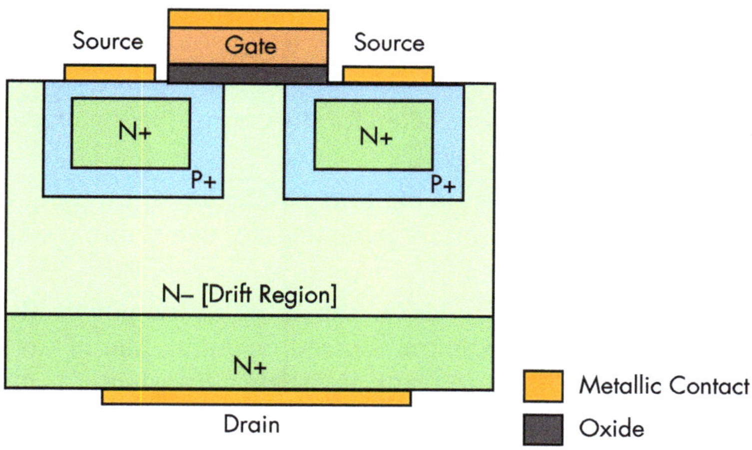


**Figure 6.4** Typical power MOSFET construction.

power MOSFET [2]. The vertical geometry of power MOSFETs allows for the drift region dimensions to be tunable during manufacturing. In comparison to a planar geometry, which would require more space on a PCB if the drift region size increases, a vertical geometry allows for more flexibility on the size of the component without the need for additional space on the PCB. Depending on the application, the drift region dimensions will vary to accommodate the power requirements.

The introduction of the N– drift region within the power MOSFET also results in the introduction of additional resistance when the MOSFET is on. This results in the requirement for a cooling infrastructure for the MOSFET to manage heat generation in applications where the MOSFET is conducting relatively higher currents or operating at relatively higher voltages. A discussion on heat sinking requirements and example cooling architectures is provided later in this chapter.

The selection of a power MOSFET for a power conversion circuit requires a full analysis of the MOSFET parameters to ensure that the selected MOSFET is appropriate for the application. Some of the parameters that must be analyzed include [3, 4]:

- Turn-on delay time ($t_{d(ON)}$): The time from when the gate-source voltage rises over 10% of $V_{GS}$ until the drain-source voltage reaches 90% of $V_{DS}$.
- Rise time ($t_r$): The time taken for the drain-source voltage to fall from 90% to 10% of $V_{DS}$.
- Turn-off delay time ($t_{d(OFF)}$): The time from when the gate-source voltage drops below 90% of $V_{GS}$ until the drain-source voltage reaches 10% of $V_{DS}$.
- Fall time ($t_f$): The time taken for the drain-source voltage to rise from 10% to 90% of $V_{DS}$.
- Switching time: Switching time is the sum of $t_{d(ON)}$, $t_r$, $t_{d(OFF)}$, and $t_f$ and is an important parameter for determining switching losses in the MOSFET.
- dv/dt capability: The dv/dt capability of the power MOSFET represents the maximum allowed changing rate of the drain-source voltage during any switching transient. A drain to source voltage rate change exceeding the maximum dv/dt rating of the power MOSFET can lead to a failure of the MOSFET.

- Body diode characteristics: A power MOSFET has a circuit structure between source and drain equivalent to a diode. The current characteristics of the body diode must be analyzed to ensure that any current flow through the body diode is not exceeded in the application.
- Total gate charge ($Q_g$): Total gate charge represents the accumulation of gate-source capacitance $Q_{gd}$ and gate-drain capacitance $Q_{gd}$ (also called the Miller capacitor). In high-frequency operations, $Q_{gs}$ should be selected as small as possible. In addition, when power MOSFETs are used in H-bridge circuits, the ratio of $Q_{gd}/Q_{gs}$ should be selected to be lower than 1 to prevent the circuit from shoot through.

Figure 6.5 [3] shows the switching time waveform of a typical MOSFET with some of its main parameter definition.

Power MOSFETs are generally most suitable for relatively lower-power DC applications with high switching frequency requirements

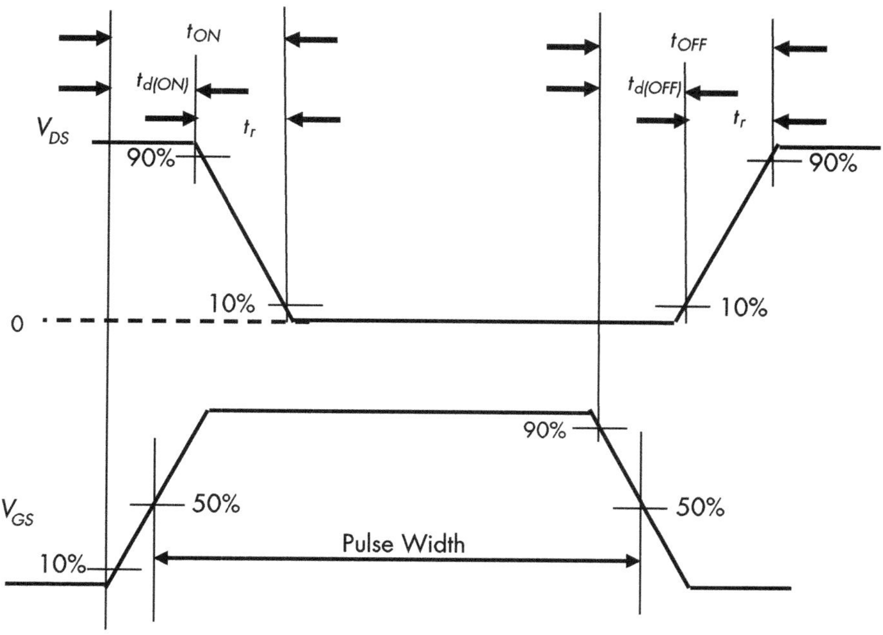


**Figure 6.5** Switching time waveform. (*After:* [3].)

(>20 kHz). A few examples of applications include DC-DC converters, small motor controllers, and power supplies [5].

## 6.3 IGBT

An IGBT is a power transistor that combines an input metal-oxide-semiconductor (MOS) and an output bipolar junction transistor. IGBTs are constructed as shown in Figure 6.6. The difference between a power MOSFET and IGBT lies in the addition of a heavily doped P-type region near the collector terminal. This heavily doped P-type region, known as the injecting layer, introduces holes to the conduction mechanism during IGBT operation. The addition of holes from the injecting layer to the drift layer allows for the comparatively higher-voltage capabilities of IGBTs. When a positive voltage bias is applied to the gate of the IGBT, a conductive channel is formed in the P-type (body) layer near the gate terminal that allows electrons to travel from the N+ regions near the emitter terminal to the collector. When bias is on, the IGBT is capable of conducting in only one direction due to the additional PN junction near the collector terminal. For this reason, IGBTs require free-wheeling diodes in parallel to allow for reverse current flow if needed in the application [6].

Figure 6.7 shows an equivalent circuit for an IGBT. When the N-channel MOSFET is turned on by applying a positive voltage to

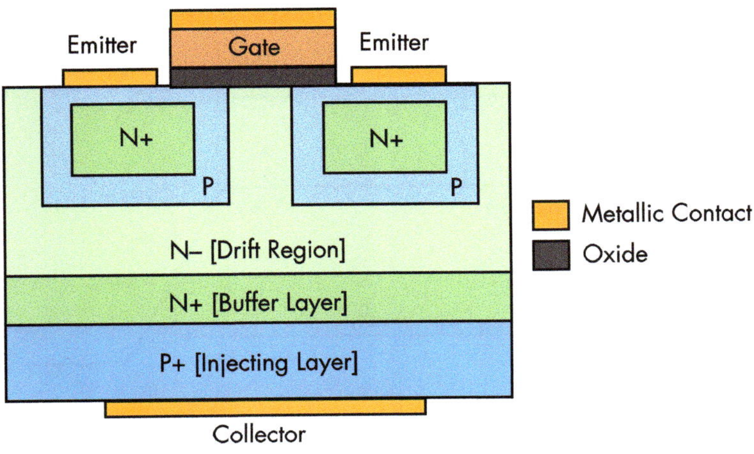


**Figure 6.6** Typical IGBT construction.

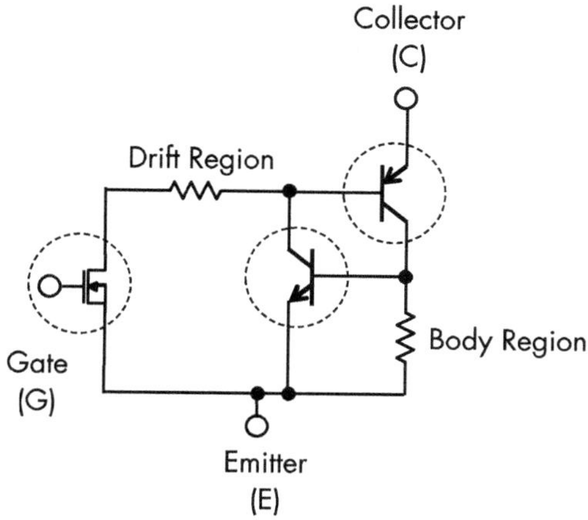


**Figure 6.7** IGBT equivalent circuit.

the MOSFET's gate, a base current flows in the PNP transistor, which turns the transistor on, resulting in current flow from the IGBT collector to the emitter. The voltage drop across the IGBT is the sum of the diode drop across the P-N junction and the voltage drop across the driving MOSFET. Thus, unlike the power MOSFET, the on-state voltage drop across an IGBT never goes below a diode threshold. The parameters that must be analyzed when selecting an IGBT for an application are very similar to the parameters for power MOSFETs.

IGBTs are most suitable for relatively high-power AC applications with relatively low switching frequency requirements (<20 kHz). IGBTs are commonly found in the fields of renewable energy inverters, uninterruptable power supplies (UPS), and induction heating.

## 6.4 THYRISTORS

Thyristors [7] are semiconducting switching devices consisting of four layers of alternating N-type and P-type material (N-P-N-P), as shown in Figure 6.8. Thyristors are composed of three terminals: anode, cathode, and gate. Thyristors have three operating modes:

1. *Forward blocking:* When a positive voltage is applied to the anode in relation to the cathode in Figure 6.8, junctions 1 and 3 become forward-biased; however, junction 2 becomes reverse-biased and thus does not allow current flow.

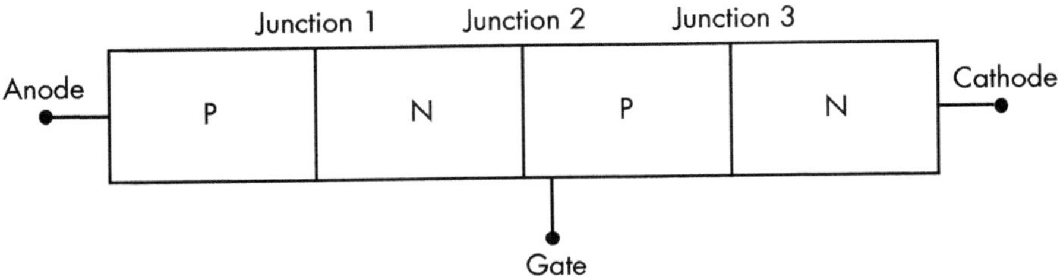


**Figure 6.8** Typical thyristor construction.

2. *Reverse blocking:* When a reverse bias is applied between the anode and cathode in Figure 6.8, junctions 1 and 3 become reverse-biased, while junction 2 becomes forward-biased and again current flow is restricted.
3. *Forward conducting:* To achieve conduction within the thyristor, a bias is applied between the gate and cathode of the device, which effectively eliminates the depletion region in junction 2, thus allowing current flow. Alternatively, a large positive voltage, above the thyristor specifications, could also be applied between the anode and cathode, effectively breaking down the depletion region in junction 2 (known as avalanche breakdown). However, this is not a typical operating mode for the thyristor, as it leads to the degradation and deterioration of the component.

Thyristors, unlike power MOSFETs and IGBTs, do not require the gate voltage to be continuously supplied for operation. Once the positive bias voltage between the gate and cathode activates, continuous current flows through the thyristor (i.e., the device is in the forward conducting mode), regardless of the magnitude of the applied gate voltage. At this point, the gate voltage can be removed and conduction through the thyristor will continue as long as the current remains above the holding current. Below the holding current, the thyristor will turn off and a bias voltage through the gate and cathode will again have to be supplied to switch the device on. Commercially available thyristor voltage and current specifications range from approximately 1 to 10 kV and upwards of 1,000A, respectively [8]. Due to these capabilities, thyristors are ideal for high-power applications including ground-fault interrupt circuits, solid state relays, high-power rectifiers, and inverters [9].

## 6.5 HEAT-SINKING REQUIREMENTS AND ARCHITECTURES

Switching components such as power MOSFETs and IGBTs can generate a large amount of heat, especially when operating at high switching frequencies due to the losses associated with the components [10–12]. This necessitates proper thermal management to ensure optimal device performance and expected lifetime of the components. The parameter of interest when evaluating heat generation and dissipation is the thermal resistance $\theta_{JA}$, which is defined as shown in (6.1):

$$\theta_{JA} = \frac{T_j - T_A}{P_D} \tag{6.1}$$

where $T_j$ is the junction temperature (maximum operating temperature of the device and it usually depends on the die used in the manufacturing process), $T_A$ is the ambient temperature, $P_D$ is the power dissipation, and $\theta_{JA}$ and $T_j$ are usually values provided within the respective component's datasheet.

Using these values and an ambient temperature $T_A$ representative of the worst-case environmental conditions that the component will encounter in the application; power dissipation values $P_D$ can be calculated that allow for the determination of the cooling infrastructure needed in the application. At a component level, most of the heat travels from the semiconductor to the copper pads underneath the component packaging that connects the component to the PCB. The copper pad serves to dissipate the generated heat through the PCB. Depending on the amount of heat that needs dissipating, the thickness of the copper can be tuned to optimize component performance.

Higher-power applications require dedicated heat sinks or other cooling mechanisms. Heat sinks are composed of a thermal conductor (usually aluminum) and are constructed with evenly spaced fins. The thermal conductor is mechanically attached to the component, usually by using screws and thermal paste, and allows for the heat to travel away from the component. The fin design allows for air to more easily flow through the heat sink for further dissipation. During installation of the heat sink, care must be taken to ensure good thermal contact between the heat sink and the component to optimize the heat transfer. Furthermore, in certain applications, fans in conjunction with heat sinks have shown to improve the heat exchange. High-power applications such as power converter circuits in EVs that consist of multiple power MOSFETs or IGBTs require liquid-cooling to dissipate the heat generated by the components during operation. The components in these applications are assembled in modules, which are connected to a liquid cooling infrastructure that dissipates the generated heat [13].

## 6.6 STEP-DOWN DC CONVERTER

A step-down DC-DC converter (more commonly called a buck converter) converts a higher input voltage to a lower output voltage. At a high level, the operating principle of a buck converter involves the controlled transfer of energy from the input to the output with the aid of switches, inductors, and capacitors. A high-side switch (often a MOSFET or IGBT)[2] and a low-side switch (a diode in the simplest manifestation of the converter) control the flow of current through the inductor and the average output voltage can be controlled by adjusting the duty cycle of the high-side switch. Figure 6.9 shows the basic buck converter circuit. When the high-side switch (MOSFET $Q$) is turned on, current flows from the input to the inductor ($L$) to charge the capacitor ($C$). Current also flows to the output load ($R_L$). When the high-side switch (MOSFET $Q$) turns off, the energy stored in the inductor is discharged through the diode. During the steady state, the circuit can operate in one of two modes depending upon the inductor current. The circuit is said to operate in the discontinuous conduction mode (DCM) if the inductor current reaches 0 and in the continuous conduction mode (CCM) if the inductor current never reaches 0. The output capacitor ($C$) smooths the output voltage to deliver a stable voltage to the load.

The circuit shown in Figure 6.9 has two limits of operation. For a pulse width modulated (PWM) duty cycle for the high-side switch of 0, the output voltage equals zero, and for a PWM duty cycle for the

2. This chapter will use a MOSFET as the switching component for discussion.

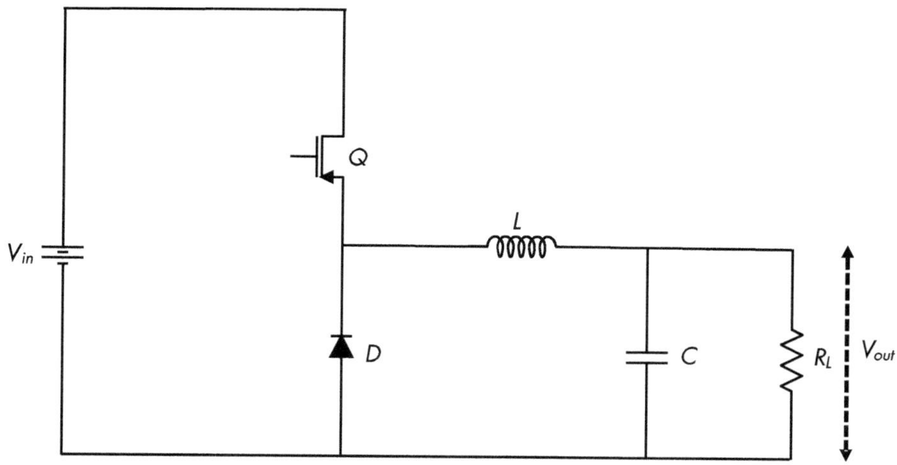


**Figure 6.9** Basic buck converter circuit.

high-side switch of 1, the output voltage equals the input voltage. In between those limits, the output voltage in CCM is given by:

$$V_{out} = D \bullet V_{in} \tag{6.2}$$

where $D$ is the PWM duty cycle for the high-side switch. The parallel combination of the inductor and capacitor in the circuit in Figure 6.9 acts as a second-order low-pass filter, which helps to reduce the voltage ripple at the output.

The circuit shown in Figure 6.9 is an example of an asynchronous buck converter. A synchronous buck converter replaces the diode in Figure 6.9 with a second MOSFET, which acts as a low-side switch. In a synchronous buck converter, both the high-side and low-side switches are driven with switching PWM pulses. The pulses for the high-side and low-side switches are alternating during each cycle. In other words, when the high-side switch is on, the low-side switch is off and vice versa. This causes the time window for the driving pulse in the synchronous converter to match the diode's forward bias window in the nonsynchronous converter. As a result, the same duty cycle calculations (6.2) used for nonsynchronous buck converters can be applied to synchronous buck converters. In other words, the system's duty cycle, which determines the step-down or step-up ratio, is the duty cycle of the high-side switch. Table 6.1 [14] provides a high-level comparison between synchronous and nonsynchronous buck converters.

In practice, switching between the high-side and low-side switches requires enforcing some dead time to prevent a momentary short-circuit that may result when the high-side switch is turned off and the low-side switch is turned on. This can occur because the high-side and low-side switches do not turn on and off instantaneously.

Although synchronous buck converters are considered to be highly efficient power converters, factors that lead to power loss that impact the converters efficiency include [15, 16]:

- *Conduction loss in switches:* The conduction losses can be estimated from the output current, the on-resistance of the MOSFETs, and the PWM duty cycle for the MOSFETs.
- *MOSFET switching losses:* This refers to the energy dissipated as heat during the transition between the on-state and off-state of a MOSFET. During MOSFET turn-on, the MOSFET gate is charged as the MOSFET switches from a nonconducting to a conducting state. The turn-on loss is determined by the gate charge and the voltage across the MOSFET during the transition. Similarly, the switching loss during MOSFET turn-off is due to the discharge of the stored charge in the MOSFET and is determined by the reverse recovery charge and the voltage across the MOSFET during the transition.
- *Reverse recovery loss in the MOSFET body diode:* This refers to the loss that occurs when the high-side MOSFET is turned on. Under this condition, the transition of the body side of the low-side MOSFET from the forward to the reverse bias state

**Table 6.1**
Synchronous Versus Nonsynchronous Buck Converters

| | **Synchronous Converter** | **Nonsynchronous Converter** |
|---|---|---|
| **Efficiency** | Typically, above 90% | Typically, around 80% |
| **Factors Determining Efficiency** | On-resistance of high-side and low-side switches and switching waveform overlap (i.e., dead time) | Diode forward voltage |
| **Ringing** | Observed at switching node | Observed at output |

*From:* [14].

generates a recovery loss in the body diode due to the diode recovery.

- *Output capacitance loss in the MOSFET:* This loss is generated during each switching cycle due to the charging of the high-side MOSFET output capacitances. The charge on the low-side MOSFETs is discharged by the inductors when the low-side MOSFETs turn on, resulting in zero voltage switching (ZVS). For this reason, the charge/discharge loss in the low-side MOSFET is usually ignored.
- *Dead time loss:* As discussed previously, a period of dead time is provided for turning off both the MOSFETs to prevent a short-circuit and a current spike from occurring when the inductor current continues to flow. During the dead time, the inductor current flows through the body diode of the low-side MOSFET.
- *Gate charge loss:* This is due to the power loss associated with charging the gate of the MOSFET and depends on the gate capacitance of the MOSFETs.
- *Inductor conduction loss:* This is due to the inductor resistance and the core loss associated with its magnetic properties.
- *Output capacitor loss:* This is due to the equivalent series resistance (ESR) of the output capacitor.

## 6.7 STEP-UP DC CONVERTER

A step-up DC-DC converter (more commonly called a boost converter) converts a lower input voltage to a higher output voltage. The basic components used in the boost converter are the same as the components used in a buck converter. The difference lies in the connection of the components. Figure 6.10 shows the basic boost converter circuit. When the switch (MOSFET $Q$) is turned on, current flows through the inductor ($L$) and the switch. When the switch is turned off, the energy stored in the inductor due to the current flow when the switch was on is released through the diode and the output load giving the load a higher voltage than the input voltage. In this way, in a boost converter, the input continuously transfers energy to the output during the switching cycle through the inductor's energy storage and

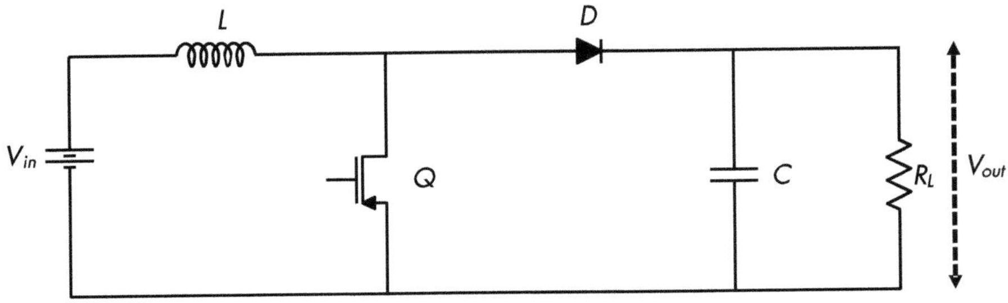


**Figure 6.10** Basic boost converter circuit.

release. Just like in the buck converter circuit, the output capacitor (*C*) smooths the output voltage to deliver a stable voltage to the load.

Also just like buck converters, boost converters can also be designed to operate in the CCM or the DCM, which differ in the behavior of the inductor current and result in different converter performance and efficiency. As discussed previously, in CCM, the inductor current remains positive and never reaches zero at any stage of the switching cycle. The inductor current grows during the switch-on time, storing energy in its magnetic field and the inductor current drops as it transfers its stored energy to the output when the switch is off. The current waveform in CCM has a triangle shape and the load current is equal to the average value of the current. CCM operation results in lower peak currents and output voltage ripples, which in turn results in higher efficiency. However, the need for continuous current flow results in greater inductor sizes, which impact the physical size of the boost converter.

Unlike CCM, the inductor current in DCM converters drops to 0 during a portion of the switching cycle. This results in the energy transfer to the output terminating before the start of the subsequent switch on period. In other words, there is a zero-current interval between the switch-off and the subsequent switch-on periods, giving the inductor current waveform a trapezoidal shape. The operation of the DCM converter results in smaller inductor values allowing for more compact converter designs. However, the use of smaller inductors results in relatively larger peak currents, increased output voltage ripple and an overall lower efficiency. For this reason, DCM converters are more frequently used in applications that have a wide range of loads or relatively lighter loads.

## 6.8 STEP-DOWN/STEP-UP DC CONVERTER

Applications where the input voltage regulation and output voltage requirements may vary require a DC-DC converter that can produce an output voltage that is either lower than or higher than the input voltage. This can be accomplished using a step-down/step-up or a buck-boost converter. A buck-boost converter operates in two distinct modes: a buck mode where the output voltage is lower than the input voltage, and a boost mode where the output voltage is higher than the input voltage. Additional circuitry is used to provide a transition between the two modes and to produce a stable output voltage when the input voltage is varying.

The buck-boost converter circuit also uses the same components as the individual buck and boost converter circuits with a different component connection scheme. Figure 6.11 shows the basic buck-boost converter circuit. Just like with the basic buck and the basic boost converter circuits discussed earlier in this chapter, the basic buck-boost converter circuit also operates using a switch (MOSFET *Q*), a diode, an inductor, and a capacitor. When the switch is on, energy is stored in the inductor, and when the switch is turned off, the energy is transferred to the output through the diode. The duty cycle of the switch determines the converter's output voltage.

When the transistor (*Q*) is turned on in Figure 6.11, the inductor (*L*) is connected to the input power source ($V_{in}$), causing the inductor current to increase linearly. The diode at this time is reverse-biased and does not conduct. When the transistor (*Q*) turns off, the inductor (*L*) current flows through the diode, hence transferring the energy stored in the inductor to the load. Due to the circuit configuration, the voltage at the load has the polarity as shown in Figure 6.11 where

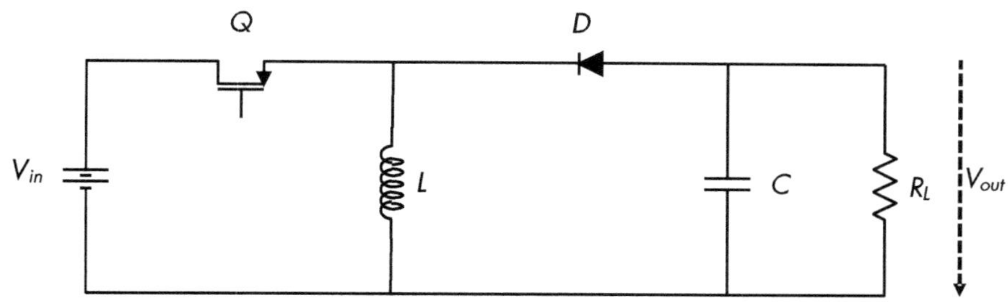


**Figure 6.11** Basic buck-boost converter circuit.

the positive terminal of the load voltage is connected to the negative terminal of the input voltage source that results in the converter performing a voltage inversion at the load.

In a steady state, the average value of the voltage across the inductor is zero, which implies that the output voltage ($V_{out}$) can be calculated as follows:

$$V_{out} = \frac{-V_{in} \bullet D}{(1-D)} \tag{6.3}$$

where $D$ is the duty cycle for the transistor ($Q$) and is in the range from 0 to 1. As can be seen in (6.3), the magnitude of the output voltage can be higher (when $D > 0.5$) or lower (when $D < 0.5$) than the input voltage. However, the output voltage will always have an inverse polarity to the input voltage [17].

The buck-boost converter circuit in Figure 6.11 is an inverting converter that may not serve the needs of applications where a positive output voltage is required. The Single-Ended Primary Inductor Converter (SEPIC), Zeta,[3] and two-switch buck-boost converter are three commonly used noninverting buck-boost topologies.

### 6.8.1 Two-Switch Buck-Boost Converters

The two-switch buck-boost converter [17] is a cascaded combination of a buck converter followed by a boost converter. As inherent in its name, the converter uses two switches for operation. Figure 6.12 shows the basic two-stage buck-boost converter configuration. The buck-boost function is realized by operating the converter in buck

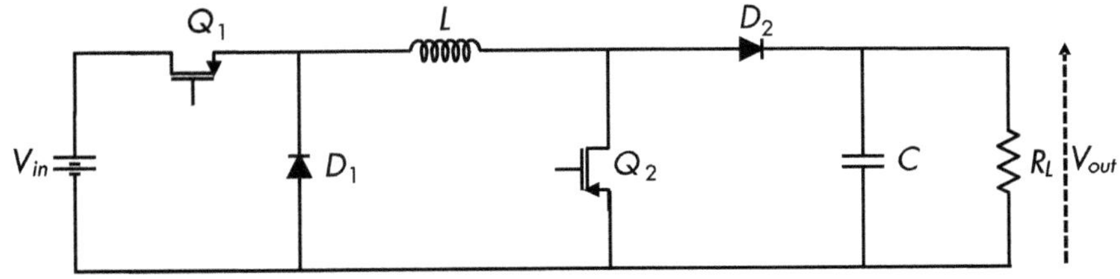


**Figure 6.12** Basic two-stage buck-boost converter circuit.

3. The Zeta converter is also sometimes called the inverse SEPIC.

mode when $V_{in}$ is greater than $V_{out}$ and in boost mode when $V_{in}$ is lower than $V_{out}$. In buck mode, MOSFET $Q_2$ is controlled to always be off, and the output voltage is regulated by controlling $Q_1$ as in a typical buck converter with the voltage conversion ratio being the same as that of a typical buck converter. In boost mode, MOSFET $Q_1$ is turned on continuously. This causes diode $D_1$ to be reverse biased allowing the converter to operate in the boost mode. Similar to the typical boost converter, the output voltage is then regulated by controlling MOSFET $Q_2$.

### 6.8.2 SEPICs

Figure 6.13 shows the basic SEPIC circuit [18]. This circuit also has a single MOSFET ($Q$) and a single diode ($D_1$) just like an inverting buck-boost converter circuit and the current and voltage requirements for these two components are similar to the requirements in the inverting buck-boost converter circuit. The SEPIC circuit uses an additional inductor ($L_2$) and an additional capacitor known as an AC-coupling capacitor ($C_p$). The SEPIC's fundamental operation relies on the energy transfer between two inductors and a capacitor. The two inductors ($L_1$ and $L_2$) and the AC-coupling capacitor ($C_p$) provide an intermediate energy storage state in the SEPIC. The output voltage is controlled by the MOSFET ($Q$) and the output capacitor ($C$). When the MOSFET is turned on, energy is stored in both inductors and the AC-coupling capacitor ($C_p$) is charged. The energy in the two inductors is transmitted to the output capacitor and the load when the MOSFET is turned off. In steady state, the average value of the voltage across the inductor is zero, which implies that the output voltage ($V_{out}$) can be calculated using the following equation:

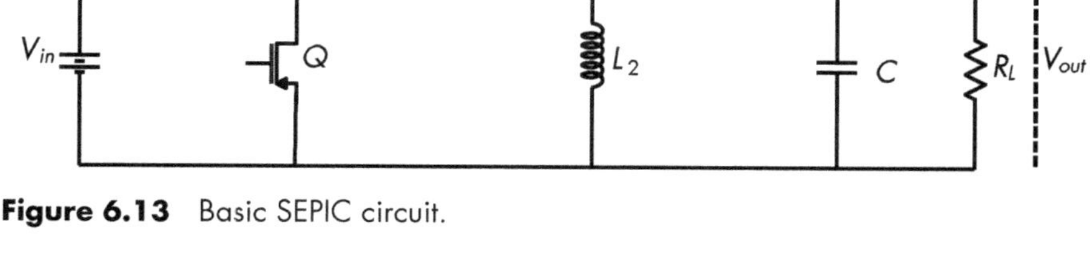


**Figure 6.13** Basic SEPIC circuit.

$$D = \frac{V_{out} + V_d}{V_{in} + V_{out} + V_d} \tag{6.4}$$

where $D$ is the duty cycle for the MOSFET ($Q$) and $V_d$ is the forward voltage drop of the diode. The coupling capacitor ($C_p$) in a SEPIC circuit sees significant root-mean-square (RMS) current relative to both the input and output current, which generates extra power loss and reduces the converter's efficiency. SEPICs are ideally suited for battery-powered devices with variable battery voltage, including for EVs, as the converter provides a constant output voltage despite a changing input voltage. The latter improves the efficiency and dependability of the entire system. These converters also have the added advantage of providing galvanic isolation for applications requiring such isolation due to the use of the AC-coupling capacitor.

### 6.8.3 Zeta Converter

Unlike a SEPIC that is configured with a standard boost converter, the Zeta converter [19] is configured from a buck converter that drives a high-side MOSFET. Figure 6.14 shows the basic Zeta converter circuit consisting of output capacitor $C$, output load $R_L$, coupled inductors $L_1$ and $L_2$, MOSFET $Q$, diode $D$, and the AC-coupling capacitor $C_p$. When MOSFET $Q$ is on, energy from the input power supply is stored in the coupled inductors $L_1$ and $L_2$ and the AC coupling capacitor $C_p$. When MOSFET $Q$ is turned off, current to the output is provided by the inductors and the stored energy in the AC-coupling capacitor. The output voltage of the circuit can be calculated as follows:

$$V_{out} = \frac{V_{in} \bullet D}{(1-D)} \tag{6.5}$$

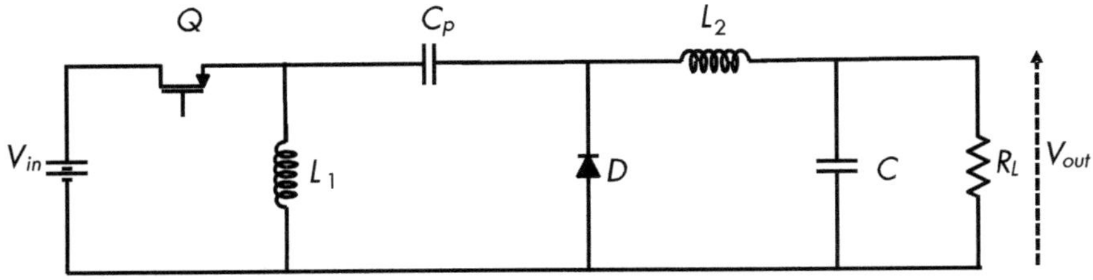


**Figure 6.14** Basic Zeta converter circuit.

where $D$ is the duty cycle for the MOSFET ($Q$).

## 6.9 COMPONENT SELECTION

The components for DC-DC converter circuits must be selected to ensure that the converter has the highest possible efficiency, power dissipation loses are minimized, and the components can handle the peak expected voltages and currents [20, 21]:

- Switching MOSFET:
  - As the power converter circuit efficiency is directly related to the switching frequency, the MOSFET should be selected with a fast-switching speed (i.e., short turn-on delay time, rise time, and turn-off delay time).
  - The gate to source cutoff voltage of the MOSFET should be lower than the input voltage to allow for a controlled start-up of the converter circuit.
  - The MOSFET's on-resistance ($R_{ds(\text{on})}$) should be as low as possible to minimize conduction losses. In practice, a trade-off is required between the on-resistance, the input capacitance, and the output capacitance.
  - For boost converters, the MOSFET's rated current should be approximately three times the peak current. However, for buck converters, a rated current of approximately three times the output current multiplied by the step-down ratio should be used when selecting a MOSFET.
  - For boost converters, the MOSFET's rated voltage should be approximately 1.5 times larger than the output voltage. For buck converters, the MOSFET's rated voltage should be approximately 1.5 times larger than the input voltage.
  - Once a MOSFET is selected for the application, the heat dissipation of the MOSFET must be analyzed to ensure that the operating temperature rating of the MOSFET is not exceeded in the application.
- Diodes:
  - Diodes are typically selected with a small forward voltage ($V_F$) to prevent losses caused by the forward voltage drop and to improve efficiency.

- The selected diode should also have a small capacitance between the diode terminals as this can impact the switching speed and the generation of noise due to the diode's turn-on and turn-off times. As discussed previously, any reduction in the switching speed results in an increased switching loss in the circuit.
- The reverse leakage current of the diode should be small to improve efficiency and reduce noise.
- For boost converters, the rated current of the diode should be approximately three times higher than the peak current of the inductor at the minimum expected input voltage. For buck converters, the rated current of the diode should be approximately three times higher than the peak inductor current at the maximum expected input voltage.
- The rated voltage of the diode should be approximately 1.5 times higher than the output voltage for boost converters and approximately 1.5 times higher than the input voltage for buck converters.

- Inductor:
  - The inductor selected must be able to handle the peak expected switching current in the circuit without saturating the inductor core as that will result in a loss of inductance.
  - The value of the selected inductor is chosen based on the desired ripple current. Most circuit designs operate with a ripple current of less than 20% of the inductor's average operating current.
  - At the expense of higher-output voltage ripple, small-value inductors result in a higher-output current slew rate, improving the load transient response of the converter. Large-value inductors lower the ripple current and reduce the core magnetic hysteresis losses.
  - The total coil losses can be combined into the loss resistance ($R_S$), which is effectively connected in series with the inductance ($L_S$). Even though the losses in $R_S$ are frequency-dependent, the DC resistance ($R_{dc}$) is also often specified in inductor datasheets. The inductor DC resistance depends on the wire material and size, and the construction type for surface mount inductors. It is characterized at room temperature

by a simple resistance measurement. The size of $R_{dc}$ directly influences temperature rises in the inductor coil. Prolonged operation above the current rating can increase the probability of an inductor failure.

- The total coil losses in an inductor consist of both the losses due to $R_{dc}$ and the following frequency dependent components:
  - Core-material losses (magnetic hysteresis loss, eddy-current loss);
  - Skin-effect losses in the conductor (current displacement at higher frequencies);
  - Magnetic-field losses of adjacent windings (proximity effect);
  - Radiation losses.
- Table 6.2 [21] provides target inductance values as a function of the oscillation frequency and output current.

## 6.10 DC-AC CONVERTERS

Most EVs on the road today use some kind of an AC motor that is driven using a H-bridge circuit. As discussed in Chapter 5, AC motors run off a sinusoidal voltage function, with the rotor following the peak of the sinusoid around the stator. To make the EV go faster, the frequency of the sinusoid must be increased. Conversely, to slow the EV down, the frequency of the sinusoid must be reduced. A DC-AC controller (commonly called an inverter) converts an input DC voltage (e.g., from a battery) to an AC voltage to drive an AC motor such as in an EV application. Inverters can be designed to operate in a single-phase or in a three-phase configuration and can be of two

**Table 6.2**
Selecting Inductance Values for a DC-DC Converter Circuit

| Load | 50 kHz | 100 kHz | 180 kHz | 300 kHz | 500 kHz |
|---|---|---|---|---|---|
| Light (<100W) | 330 μH | 220 μH | 100 μH | 47 μH | 22 μH |
| Medium (100–500W) | 220 μH | 100 μH | 47 μH | 22 μH | 10 μH |
| Heavy (>500W) | 100 μH | 47 μH | 22 μH | 10 μH | 6.8 μH |

*From:* [21].

types. Voltage source inverters (VSIs) produce the desired AC output voltage, while the input DC voltage remains constant. Current source inverters (CSIs) use a constant input current to produce the desired output AC voltage.

Figure 6.15 shows the basic architecture of a single-phase, half-bridge inverter, which consists of two switching MOSFETs ($Q_1$ and $Q_2$) linked in series with a DC voltage source, two feedback diodes ($D_1$ and $D_2$), and two capacitors ($C_1$ and $C_2$) that link the source and the load ($R_L$). The load in this configuration is connected between the midpoint of the capacitors and the midpoint of the diodes and switching transistors. The half-bridge inverter has two modes of operation. The first mode involves turning MOSFET $Q_1$ on and MOSFET $Q_2$ off. When MOSFET $Q_1$ is on, current flows through the load ($R_L$) and MOSFET $Q_1$ and the voltage across the load ($R_L$) is equal to half the input voltage ($V_{in}$). In the second mode, MOSFET $Q_2$ is on and MOSFET $Q_1$ is off. In this mode, current flows through the load and MOSFET $Q_2$ and the output voltage is equal to the negative half of the input voltage. For inductive loads, the half-bridge inverter has four operating modes and the feedback diodes ($D_1$ and $D_2$) play a critical role when the MOSFETs are off as they provide a route for the inductive load current to return to the source. This protects against the generation of voltage transients and ensures a steady current across the load during switching transitions.

Figure 6.16 shows a H-bridge arrangement of four MOSFETs and four feedback diodes that are used in a full-bridge inverter topology. In comparison to the half-bridge architecture, this architecture provides a larger output voltage capability. Just like the half-bridge inverter, the

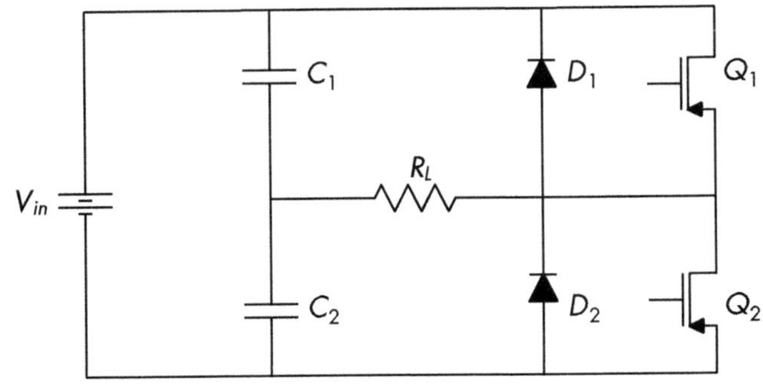


**Figure 6.15** Half-bridge inverter circuit.

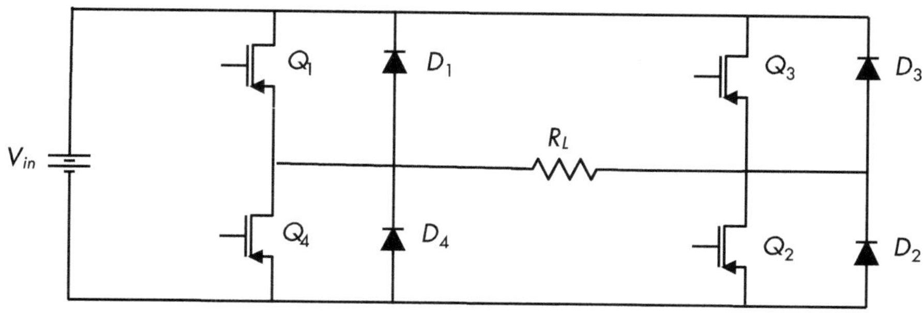


**Figure 6.16** Full-bridge inverter architecture.

full-bridge inverter also has two modes of operation. In the first mode of operation, MOSFETs $Q_1$ and $Q_2$ are turned on, while MOSFETs $Q_3$ and $Q_4$ are off, and the load voltage is equal to the input voltage. In the second mode of operation, MOSFETs $Q_3$ and $Q_4$ are turned on, while MOSFETs $Q_1$ and $Q_2$ are off, and the load voltage is equal to the negative of the input voltage. Just like with the half-bridge inverter design, the four free-wheeling diodes in the full-bridge inverter design enable the continuous flow of current through the load during the switching transitions and protect against the generation of voltage transients due to the inductive load. In order to approximate the actual shape of a sinusoid, PWM is used in the full-bridge circuit for switching the MOSFETs. Short pulses are used to approximate low voltages and long pulses are used to approximate high voltages. The use of PWM results in a waveform that creates a sawtooth waveform, which roughly approximates a sine wave. By varying the width of the pulses, a sinusoid of any frequency can be approximated which allows for change in, for example, the speed of an AC motor. The H-bridge circuit shown in Figure 6.16 has no ability to modulate the amplitude of the voltage. Modulating the amplitude of the voltage would require a DC-DC controller in series with this circuit.

Figure 6.16 shows an example of a single-phase inverter. Most EV AC motors are three-phase motors that require three-phase inverters. Figure 6.17 shows the architecture of a three-phase inverter. The 120° conduction mode and the 180° conduction mode are the two fundamental operating modes for three-phase full-bridge inverters. In the 120° conduction mode, within each phase leg, there are only two MOSFETs that are turned on at any time and each MOSFET conducts

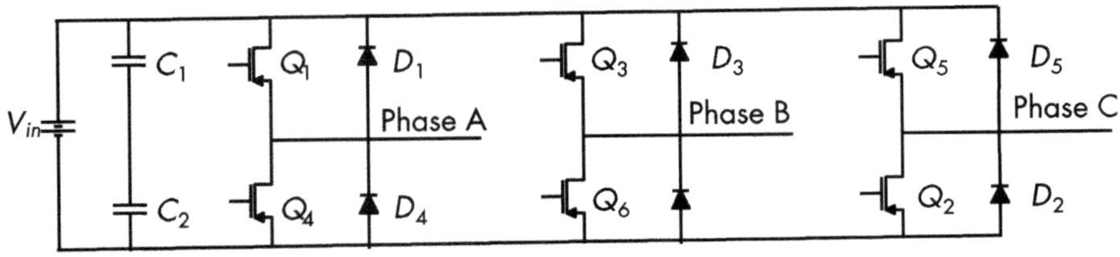


**Figure 6.17** Three-phase inverter architecture.

for only one-third of the fundamental period (360°). In the 180° conduction mode, three MOSFETs are on at any time and each MOSFET conducts for 180° of the fundamental output voltage waveform. Just as in the half-bridge and full-bridge inverter architectures, the freewheeling diodes in the three-phase inverter circuit play an important role for inductive loads by enabling the continuous flow of current through the load during the switching transitions and by protecting against the generation of voltage transients due to the inductive load. The 120° conduction mode of each transistor results in underutilization when compared to the 180° conduction mode for the identical load state. Due to these reasons, three-phase inverters are often operated in the 180° conduction mode.

## References

[1] Pressman, A., "MOSFET Power Transistors and Input Drive Circuits," in S. Chapman, (ed.), *Switching Power Supply Design*, Second Edition, New York: McGraw-Hill, 1998, pp. 353–380.

[2] Monolithic Power Systems, "MOSFET Switches: Basics and Applications in Power Converters," Article #A-0027 Rev. 1.0, April 1, 2022.

[3] Toshiba, "Power MOSFET Electrical Characteristics," Application Notes, January 10, 2023.

[4] Texas Instruments, "Fundamentals of MOSFET and IGBT Gate Driver Circuits," Application Report, March 2017.

[5] https://www.infineon.com/cms/en/product/power/mosfet.

[6] "ON Semiconductor: Motor Control IGBTs and Free-Wheeling Diodes," AND9088/D - Rev 3, August 2016, https://www.onsemi.jp/pub/Collateral/AND9088-D.PDF.

[7] Sze, S. M., "Thyristors," in *Semiconductor Devices: Physics and Technology*, Third Edition, New York: Wiley, 2012, pp. 123–159.

[8] https://www.mitsubishielectric.com/semiconductors/powerdevices/products/high-power/thyristor/.

[9] STMicroelectronics, "Basics on the Thyristor (SCR) Structure and Its Application," Application Note, AN4607 – Rev 2, September 2018.

[10] Selgneurin, A., TN1378 Technical Note, "HU3PAK Package Mounting and Thermal Behavior," STMicroelectronics, 2021.

[11] Texas Instruments, *Thermal Design by Insight, Not Hindsight,* Application Report, AN-2020, SNVA419C, revised April 2013.

[12] Texas Instruments, *Understanding Thermal Dissipation and Design of Heatsink,* Application Report, SLVA462, May 2011.

[13] Moreno, G., et al., "Electric-Drive Vehicle Power Electronics Thermal Management: Current Status, Challenges, and Future Directions," *Journal of Electronic Packaging,* Vol. 144, No. 1, March 2022, p. 011004.

[14] https://resources.pcb.cadence.com/blog/synchronous-vs-nonsynchronous-dc-dc-conversion.

[15] ROHM Semiconductor, "Efficiency of Buck Converter," Application Note, No. 64AN035E Rev.004, November 2022.

[16] Renesas, "SLG47105, Buck, Boost, Buck-Boost Converter," Application Note, R19AN0248EE0100, Rev 1.00, 2023.

[17] Fan, H., Texas Instruments, "Design Tips for an Efficient Non-Inverting Buck-Boost Converter," *Analog Applications Journal,* AAJ 3Q, 2014.

[18] Texas Instruments, "AN-1484, Designing A SEPIC Converter," Application Report, SNVA168E, May 2006, revised April 2013.

[19] Falin, J., Texas Instruments, "Designing DC/DC Converters Based on ZETA Topology," *Analog Applications Journal,* 2Q 2010.

[20] Texas Instruments, "Choosing Inductors and Capacitors for DC/DC Converters," Application Report, SLVA157, February 2004.

[21] Torex Semiconductor, Ltd, "How to Select External Components for DC/DC Converters," Technical Information Paper, 2015.

# 7

# ADVANCED DRIVER ASSISTANCE SYSTEMS

Advanced driver assistance systems (ADAS) are electronic systems in a vehicle that use advanced technologies to assist the driver. ADAS use sensors in the vehicle such as radar and cameras to perceive the world around it, and then either provides information to the driver or takes automatic action based on what it perceives. Examples of ADAS include forward collision warning (FCW), blind-spot monitoring, rear backup warning, and automatic emergency braking (AEB). ADAS have not been developed specifically for EVs and are found in a large population of conventional gasoline vehicles. Most if not all EVs use a number of different ADAS features, making their discussion in this book especially relevant as EVs are often considered more technologically advanced than their gasoline-powered counterparts. Table 7.1 [1] shows a timeline for the evolution of the many vehicle control systems incorporated in both conventional gasoline vehicles and EVs including ADAS.

## 7.1 LEVELS OF AUTOMATION

The Society of Automotive Engineers (SAE) has developed a widely used classification system that defines the degree of driving automation a car and its equipment may offer, SAE J3016 [2]. This classifica-

**Table 7.1**
Timeline for the Evolution of Vehicle Control Systems Including ADAS

| Timeline | Function | Features |
|---|---|---|
| 1980 onwards | Vehicle Dynamic Stabilization | Antilock Braking System (ABS) |
| | | Traction Control System |
| 1990 onwards | Vehicle Dynamic Stabilization | Electronic Stability Control |
| | Driver Warning and Comfort | Navigation System |
| | | Park Assist |
| 2000 onwards | Driver Warning and Comfort | Lane Departure |
| | | Adaptive Cruise Control |
| | | Night Vision |
| 2010 onwards | Automated Driving Systems | Collision Mitigation |
| | | Collision Avoidance |
| 2015 onwards | Automated Driving Systems | Low-Speed Automated Driving |

*From:* [1].

tion system, which was first introduced in 2014 (also adopted by the National Highway Traffic Safety Administration (NHTSA)), ranges from levels 0 to 5 with the driving automation spectrum beginning with vehicles without any driving automation technology and ending with entirely self-driving vehicles.

SAE J3016 refers to three primary actors in the context of ADAS:

1. The driver;
2. The driving automation system;
3. Other vehicle systems and components that do not include the driving automation system but may share hardware and software components with the driving automation system.

The Dynamic Driving Task (DDT) involves the real-time operational and tactical functions to operate a vehicle, which include:

- Lateral vehicle motion via steering;
- Longitudinal vehicle motion via acceleration and deceleration;
- Monitoring the driving environment via object and event detection, recognition, and response preparation;
- Object and event response execution.

The DDT fallback involves the response by the driver to either perform the DDT or achieve a minimal risk condition[1] after the occurrence of a DDT performance-related system failure. The levels of driving automation are defined using the specific role played by each of the three primary actors in performing the DDT and the DDT fallback.

- *Level 0: No Driving Automation:* The driver is in charge of the entire DDT and of operating the vehicle's movement, including steering, braking, parking, and any other necessary maneuver to move the car in any direction. At level 0, the driver support system may temporarily intervene during driving. Examples include stability control, FCW, AEB, and blind-spot monitoring. These technologies are considered Level 0 because they do not drive the vehicle but offer alerts or momentary action in specific situations.
- *Level 1: Driver Assistance:* At this level, sustained and operational design domain (ODD)[2]-specific execution is performed by the driving automation system of either the lateral or the longitudinal vehicle motion (but not both simultaneously). In other words, the vehicle has at least one driver support system that provides steering assistance or braking and acceleration assistance. However, the driver remains responsible for driving the vehicle and must be prepared to take control at any time and for any reason. Adaptive cruise control and lane-centering assistance are examples of a Level 1 driving assistance technology.
- *Level 2: Partial Driving Automation:* Applies to vehicles with ADAS that can take over steering, acceleration, and braking in specific scenarios. Even though the Level 2 driving support can control the primary driving task, the driver must remain

1. SAE J3016 defined the minimal risk condition as a "stable, stopped condition to which a user many bring a vehicle after performing the DDT fallback in order to reduce the risk of a crash when a given trip cannot or should not be continued."
2. SAE J3016 defined the operational design domain as the "operating conditions under which a given driving automation system or feature thereof is specifically designed to function, including, but not limited to, environmental, geographical, and time-of-day restrictions, and/or the requisite presence or absence of certain traffic or roadway characteristics."

alert and is required to actively supervise the technology at all times.

- *Level 3: Conditional Driving Automation:* Applies to vehicles that use various driver assistance systems and artificial intelligence to make decisions based on changing driving situations around the vehicle. Mercedes-Benz plans to launch a Level 3 system (called Drive Pilot) with its S-class and EQS model vehicles. These Level 3 systems will only work in limited situations when initially released and only at speeds up to 40 mph in the state of Nevada and California [3].
- *Level 4: High Driving Automation:* Applies to vehicles where the sustained and ODD-specific operation of the entire DDT and the DDT fallback is performed by the driving automation system. The capability of the driving automation system of performing the DDT fallback operation and achieving a minimal risk condition is the main difference between Level 4 and a Level 3 system. An example of a Level 4 system currently in the field is from Waymo, an Alphabet subsidiary, that introduced a Level 4 self-driving taxi service in Phoenix, Arizona, in 2023 without a safety driver in the seat [4].
- *Level 5: Full Driving Automation:* Applies to vehicles with a driving automation system that can operate the vehicle on-road anywhere within its region of the world and under all road conditions in which a conventional vehicle can be reasonably operated by a typically skilled human driver.

Table 7.2 [2] summarizes the role of the driver under different driving automation levels in a vehicle.

## 7.2 BENEFITS OF ADAS

NHTSA reported that in 2021, 42,939 people lost their lives to motor vehicle crashes, many of which were tied to human error [5]. With those statistics in mind, the main goal of ADAS is to reduce and/or prevent accidents by minimizing the amount of human interaction, and thus human error, needed throughout the DDT through increased automation. Various research organizations and federal entities in partnership with automobile manufacturers have performed detailed

**Table 7.2**
Driving Automation Levels

| | Level 0 | Level 1 | Level 2 | Level 3 | Level 4 | Level 5 |
|---|---|---|---|---|---|---|
| **Who Drives** | Drive the vehicle regardless of system engagement status | | | Driver is not driving when the feature is engaged | | |
| **Driver Responsibility** | Constantly supervise the features and steer, brake, and accelerate as needed to maintain safety | | | Driver must drive when requested by feature | Systems do not require driver to take over driving | |
| **Feature** | Provide warning and limited assistance | Provide steering OR brake/acceleration support | Provide steering AND brake/acceleration support | To take over driving when requested by feature | | |
| **Example** | AEB<br>Blind-spot warning<br>Lane departure warning | Lane centering OR adaptive cruise control | Lane centering AND adaptive cruise control | Traffic jam assist | Local driverless taxi | Same as level 4 feature but can drive anywhere under all conditions |

*From:* [2].

safety analysis validating the safety benefits of ADAS when integrating these systems in consumer vehicles. The Partnership for Analytics Research in Traffic Safety (PARTS) reported on the real-world effectiveness of ADAS in passenger vehicles [6] and found that FCW and AEB together lead to a reduction of approximately 50% in all front-to-rear crashes when compared against vehicles not equipped with FCW or AEB. The study further found that vehicles equipped with lane departure warning (LDW), lane-keeping assist (LKA), and lane-centering assist (LCA) systems are effective at reducing single-vehicle road-departure crashes when compared with vehicles with none of these lane management features. The Insurance Institute for Highway Safety (IIHS) and the High Loss Data Institute (HLDI) have also reported on the benefits of various crash avoidance technologies [7] including:

- Blind-spot detection reducing lane-change crashes with injuries by approximately 23%.
- Rear automatic braking (when combined with rearview cameras and parking sensors), reducing backing crashes by approximately 78%.

- AEB reducing large truck front-to-rear crashes by approximately 41%.

Additionally, ADAS seek to protect not only the driver but also the pedestrian. A study from IIHS [8] investigated the effects of AEB on pedestrian crash risk and found pedestrian automatic emergency braking (PAEB) resulted in a statistically significant reduction of approximately 25% to 27% in pedestrian crash risk and 29% to 30% in pedestrian injury crash risk. Through the latter IIHS study, key limitations of PAEB were also identified including its ineffectiveness in dark conditions, at speeds of 50 mph or greater, or while the vehicle is turning. These findings supported IIHS's petition to NHTSA to mandate nighttime testing to NCAP (New Car Assessment Program) [9]. As described, vehicle automation promises increased safety to drivers and pedestrians. Looking beyond safety, NHTSA also describes ADAS as providing increased mobility, economic and societal benefits, environmental benefits, and efficiency and convenience [10].

In terms of mobility, ADAS technologies allow for individuals with a disability or limited mobility to drive more confidently due to features such as ACC, LKA, and AEB. Furthermore, Level 4 and Level 5 vehicles have the potential to provide transportation options, such as through driverless taxis, for individuals not capable of performing a driving task or that are not comfortable doing so.

A NHTSA study [11] reported that in 2019 vehicle crashes within the United States resulted in a total cost of $340 billion. When quality-of-life valuations were considered, the total economic and societal harm in 2019 came to a total of $1.4 trillion. Some of the adverse driver behaviors that were identified through the 2019 NHTSA study included driver distraction, speed, and alcohol involvement. ADAS features, as well as vehicular technological advances in general, including improved driver monitoring, LKA and ACC, to name a few, aim to target these adverse situations and bring the vehicle to a safe state either by alerting the driver, taking evasive action, and/or slowing down or bringing the vehicle to a stop. Such mitigation strategies would greatly lessen the economic and societal costs by ultimately reducing the number of crashes on the road.

Additionally, due to the alarming and evident signs of climate change, the transportation sector is steadily moving toward automation and electrification, both promising a reduction in emissions.

Through features such as ACC and LCA and advanced navigation algorithms, more efficient driving, minimizing unnecessary braking or accelerating, and higher fuel efficiency (for internal combustion engine vehicles) can be achieved. The potential for automated ride share could also drastically reduce pollutants and transform land use.

## 7.3 ARCHITECTURE

Any modern vehicle including EVs is typically equipped with a number of vehicle electronic control modules that perform dedicated functions in the vehicle. As an example, in a conventional gasoline vehicle, when the driver depresses the accelerator pedal to move the vehicle, a vehicle electronic control module called the engine control module (ECM) detects the amount of accelerator pedal depression and controls the amount of fuel and air introduced into the engine to make the vehicle move. Sensors on the accelerator pedal provide accelerator pedal depression information to the ECM and, in turn, the ECM drives a motor and a fuel pump to control the amount of air and fuel introduced into the engine. Similarly, other vehicle electronic control systems such as the electronic power-assisted steering (EPAS) system, and the ABS also use information from one or more sensors to control a series of actuators to perform their function.

Just like any vehicle electronic control system, ADAS also typically control one or more outputs based on information received from one or more sensors located in the vehicle. As an example, FCW and AEB systems rely on information from one or more cameras and a radar to detect objects in front of and around the vehicle. The camera(s) tell the system what is in front of the vehicle and the radar tells the system how far it is in front of the vehicle. This information together with information from other vehicle electronic control systems or modules in the vehicle (e.g., vehicle speed information from the ABS module, steering angle, and steering rate information from the EPAS system) is used to calculate the probability of a collision. The system is designed to alert the driver (typically through a chime and a message on the vehicle's dash) and activate the brakes if a collision is deemed unavoidable.

ADAS are using an increasing number of sensors to perform their functions. These include cameras, medium and long-range radar, ultrasonic sensors, and LiDAR. A fusion process is used to authenticate the data that these sensors generate during operation to enable the software to perform the necessary tasks to active the ADAS to take the desired action. Figure 7.1 shows a high-level ADAS architecture. As can be seen in Figure 7.1, ADAS typically consist of the following [12]:

- A sensor array where one or more sensors gather information about the environment;
- A perception layer where the information from the sensor array is fused and prepared for the decision-making layer;
- A decision-making layer that uses the information from the perception layer to decide when an action should be taken;
- An action layer that is controlled by the perception layer.

### 7.3.1 Sensor Calibration

Sensor calibration is the foundational block of any ADAS and must be performed correctly before sensor fusion and obstacle detection processes may be implemented. Sensor calibration is vital for the correct operation of the ADAS as it provides the system with information about the sensors' position and orientation in real-world coordinates by comparing the relative positions of known features as detected by the sensors. Precise calibrations are also needed for further processing steps such as sensor fusion and implementation of algorithms for obstacle detection. The three categories of sensor calibrations include [13]:

- Intrinsic calibration addresses sensor-specific parameters and is conducted before extrinsic calibration. This calibration esti-

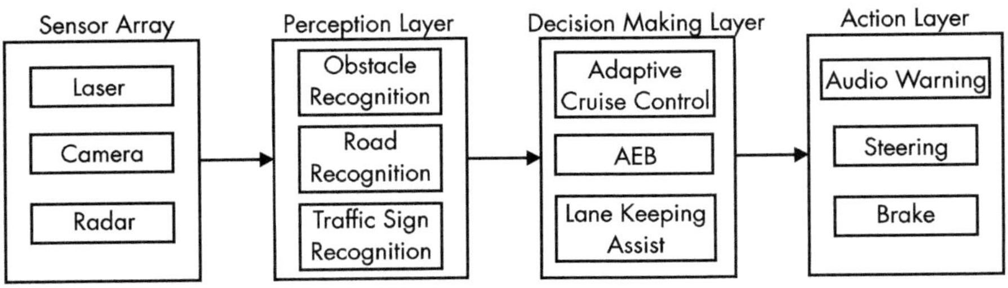


**Figure 7.1** High-level ADAS architecture.

mates the internal or intrinsic parameters of a sensor. As an example, intrinsic calibration is used to estimate the focal length of a camera. These calibrations are assumed to be constant once determined.

- Extrinsic calibration that estimates the position and orientation of a sensor relative to the three orthogonal axes of 3D space with respect to an external frame of reference. The calibration process outputs the extrinsic parameters that are consistent with the rotation and translation information of the sensor and is commonly represented in a $3 \times 4$ matrix. There are many methodologies that can be used for extrinsic calibrations. As an example, a target-based extrinsic calibration approach uses specially designed calibration targets, such as marker-less planar patterns, and checkboard patterns, to calibrate multiple sensors modalities.
- Temporal calibration that determines the synchronicity (relative time delay) of multiple data streams with different frequencies and latencies in a multisensor setup. As an example, the data capture rate for a camera at 30 frames per second may be different from the capture rate of a LiDAR operating at 5 Hz. Temporal calibration can be performed either through external synchronization where an external hardware clock that is precisely related to a real-time standard for time synchronization or internal synchronization that uses the timestamps is attached to each sensor's data to obtain a consistent view of time across all sensor networks.

### 7.3.2 Sensor Fusion

Sensor fusion and perception are at the heart of ADAS. Together, sensor fusion and perception enable the vehicle to understand the environment in which it is operating. If sensor data is not fused, the system may get contradicting inputs from sensors and be unable to determine with a degree of certainty the next course of action. Sensor fusion is the art of harmonizing data from various sensors into a unified, coherent representation of the surroundings. This integration is essential because each sensor excels in specific conditions and environments. For instance, a camera provides high-resolution visual data, radar performs well in adverse conditions and LiDAR offers precision in distance measurement. An electronic control unit (ECU) processes incoming sensor data, blends it, and presents a holistic view of the vehicle's surroundings. Sensor fusion can be done in one of three different configurations [14]:

- Complementary sensor fusion where data from each sensor is combined together to provide the ECU with the information about the environment. As an example, multiple radars on various sides of a vehicle can combine to warn the driver of an object in the vehicle's immediate vicinity.
- Competitive sensor fusion where each sensor delivers the same measurement and the system processes all the data to get a reliable reading of the environment. An inherent advantage of this configuration is the redundancy provided as the malfunction of a sensor does not impact system performance due to other sensors providing the same information.
- Cooperative sensor fusion where two sensors provide information that would not be available using only one sensor. An example would be using multiple cameras with different specifications for 3D imaging and depth understanding, which would not be possible by the use of a single camera.

### 7.3.3 Neural Networks

Neural networks are at the heart of how the perception layer in ADAS performs its function. At a high level, neural networks can be characterized as computational models with particular properties such as the ability to adapt or learn, to generate, or to cluster or organize data. The operation of neural networks is based on parallel processing. The architecture of each network is based on similar building blocks that perform the processing. The network consists of a pool of simple processing units that communicate by sending signals to each other over a large number of weighted connections. A neural network typically has the following components [15]:

- A set of processing units called neurons or cells together with a state of activation for each unit that is equivalent to the unit's output. The processing units perform a relatively simple function of receiving inputs from neighbors or external sources and using this to compute an output signal that is propagated to other units. In a typical neural network, many processing units carry out the computations in parallel.

- Connection between the units with each connection defined by a weight. The weight indicates the effect that the signal has on the unit with which it connects. Positive contributions are considered as an excitation, while negative contributions are considered an inhabitation.
- A propagation that determines the effective output of a unit from its external inputs.
- An activation function that determines the level of activation based on the effective input and the current activation.
- A method for information gathering also called the learning rule.
- An environment within which the system must operate, providing input signals, and, if necessary, error signals.

Each processing unit in a neural network performs a relatively simple task of receiving inputs from neighbors or external sources and using these inputs to compute an output signal that is propagated to other neurons. The output signal is computed by an activation function. Once the output is generated from the final neural net layer, a loss function (i.e., input versus output) is calculated and back propagation is performed where the weights are adjusted to minimize the loss function. The overall goal is to find the optimal weight values. Figure 7.2 shows an example of an artificial neuron.

#### *7.3.3.1 Terminology*

The following is terminology used for neural networks:

- *Neuron:* A basic building block of a neural network that takes weight values, performs a mathematical calculation, and produces an output. It is also called a unit, node, or perceptron.
- *Inputs:* Data/value passed to the neuron.
- *Weights:* Machine-learned numeric values that are multiplied by inputs and modified to minimize the loss function.

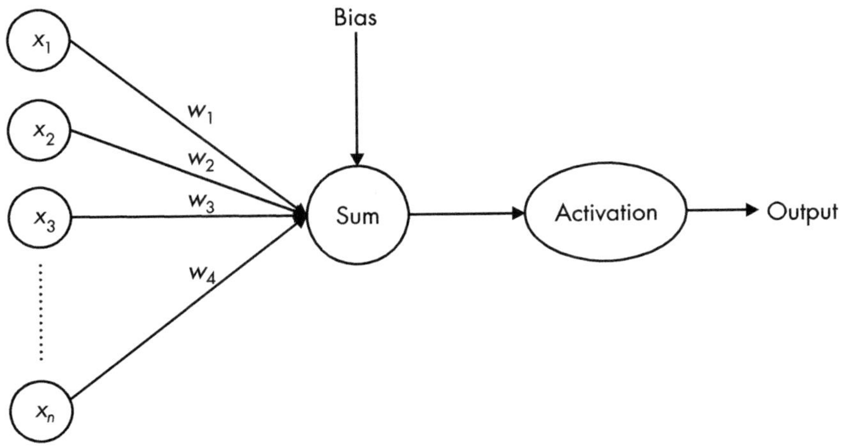


**Figure 7.2** An artificial neuron. (*After:* [16].)

- *Bias:* A constant value added to the sum of the product between input values and respective weights. It is used to accelerate or delay the activation of a given node.
- *Activation function:* A mathematical formula that generates an output from a neuron by processing the weighted inputs.
- *Output:* The activation value of a neuron (i.e., the threshold value above which the neuron is activated and produces 1 as its output).
- *Input layer:* The first layer of neurons in an artificial neural network. This layer receives input data from the outside world.
- *Output layer:* Represents the final layer of the neural network where the desired prediction is obtained.
- *Hidden layer:* The layer between the input and output layer. It is this layer that makes a neural network "deep." The primary role of the hidden layer is to transform the inputs into something that can be used by the output layer.

#### *7.3.3.2 Types of Neural Networks*

Neural networks come in many different forms and can be classified based on their structure, dataflow, neurons used, layers, and their depths of activation filters. Figure 7.3 shows an example of a simple neural network that contains a single hidden layer.

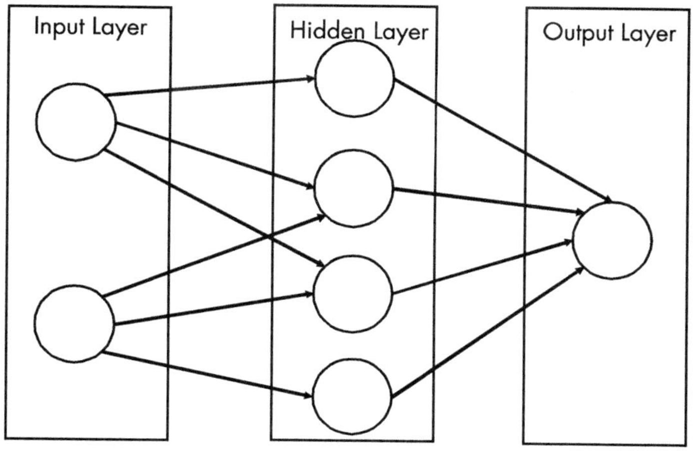


**Figure 7.3** A simple neural network.

Examples include [15]:

- *Feedforward neural networks:* Input data travels in one direction passing through artificial neural nodes and exiting through output nodes. The data processing can extend over multiple layers but no feedback connections are present. The network uses static weights for the nodes and an activation function that is typically a step function is fed by inputs, which are multiplied by weights. The neuron is activated if it is above a threshold and produces 1 as an output. Feedforward networks are simple to maintain and are equipped to deal with data, which contains a lot of noise. These networks are often used for applications such as face recognition, computer vision where target classes may be difficult to classify, and speech recognition. Figure 7.3 is an example of a feedforward network with a single hidden layer.
- *Recurrent neural networks:* These pick up where feedforward networks leave off. These networks reprocess their own outputs to generate more accurate feature outputs creating a feedback loop (i.e., a recurring process) in the hidden layer. In a recurrent network, the first layer is typically a feedforward neural network followed by a recurrent neural network layer where some information in the previous time step is remembered by

a memory function for future use. If the prediction is incorrect, the learning rate is employed to make small changes, making it gradually increase towards making the right prediction during the backpropagation. Applications of recurrent networks include text processing such as auto suggest, grammar checks, text-to-speech processing, and translation.

- *Convolutional neural networks:* These are often used for pattern recognition, image recognition, and image classification applications. These networks contain 3D arrangements of neurons instead of a standard 2D array. A convolutional neural network typically has three layers: a convolutional layer, a pooling layer, and a fully connected layer:
  - The convolutional layer performs the main portion of the network's computation load. This layer applies a set of learnable filters (also known as kernels) to the input data. Each filter extracts specific features from the input by sliding over the entire image, capturing patterns and spatial hierarchies.
  - The pooling layer reduces the spatial dimensions of the convolved feature maps to decrease computation but maintain the most important information.
  - The network typically ends with one or more fully connected layers that perform classification based on the learned features.
- *Modular neural network:* A modular neural network is made up of several neural network models that are linked together via an intermediary. Modular neural networks allow for more complex management and handling of more basic neural network systems. The multiple neural networks act as modules, each solving a portion of the problem. An integrator is responsible for dividing the problem into multiple modules as well as integrating the answers of the modules to create the system's final output. The modular neural network allows for the performance of a large and complex computational process to be done significantly faster as it is broken down into independent components. The computation speed increases because the networks are not interacting with or even connected to each other. Examples of applications include stock market prediction systems.

- *Radial basis function network:* These networks only consist of an input layer, a single hidden layer, and an output layer. The input layer simply feeds data to the hidden layer. The hidden layer of the network transforms the input data into a higher-dimensional space using radial basis functions. Radial basis functions are centered at specific points in the input space and compute similarity or distance between the input data and a center (representative) point. The most common used radial basis function is a Gaussian function.

## 7.4 SENSORS

As previously discussed, ADAS use various sensors, including cameras, radar, LiDAR, and ultrasonic sensors, to be able to make sense of the world around them and either warn the driver of an imminent hazard or take evasive action if the hazard is deemed unavoidable as shown in Figure 7.4 [17]. This section will provide an overview of the operation of some commonly used ADAS sensors.

### 7.4.1 Camera Sensors

Advanced digital cameras pointing in various directions are typically used in ADAS applications to aid the system in building a 3D model of the surrounding environment. These cameras are used to inform several ADAS subsystems by tracking lane markings, recognizing stop

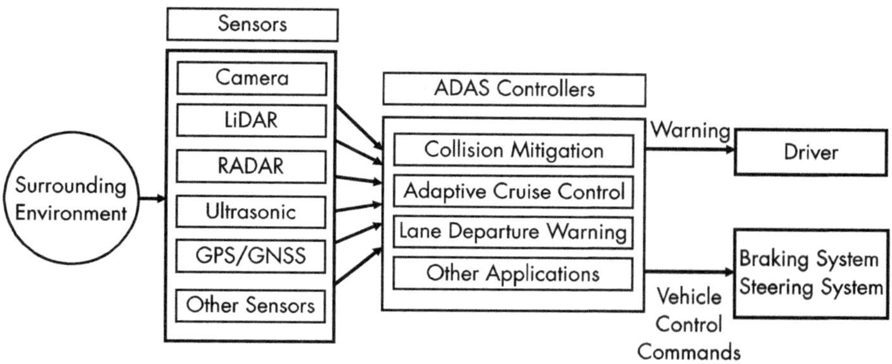


**Figure 7.4** ADAS sensor architecture. (*After:* [17].)

signs, traffic lights, and pedestrians, and detecting obstructions in the path of the vehicle. Once the images are captured and processed by the camera, the visual information is transmitted to the camera ECU, which then analyzes the data using sophisticated algorithms and communicates this information to ADAS subsystems such as AEB, LDW, LKA, pedestrian detection, traffic sign recognition, and parking assist. As described, the camera system consists of two main subsystems [18]:

- The *camera module,* which consists of lenses and an image sensor assembly. The image sensor is used to convert images perceived by the lenses into electrical signals for internal processing. Like how a human eye functions, image recognition for cameras in ADAS applications is achieved through the capture of light. The light particles known as photons are captured by using a silicon lattice, consisting of millions of pixels, which make up the image sensor. When a silicon atom or pixel site absorbs a photon, an electron is released. As increasingly more photons are captured by the image sensor, a flow of electrons is created. The job of the image sensor at this point is to use this electron flow and convert it into an electrical signal or voltage that can be read by the camera ECU. Within ADAS applications, there are two types of image sensors that are commonly used: charge-coupled device (CCD) and complementary metal-oxide semiconductor (CMOS). The difference between the two sensors is the mode by which they perform the electron to voltage conversion:
  - For CCD sensors, once photons are absorbed and converted into electrons, these electrons are passed down sequentially from row to row of the pixel array. Once the electrons are transferred to the last row of the pixel array, the electron to voltage conversion occurs. Succinctly, the whole sensor array shares one electron to voltage conversion circuit.
  - For CMOS sensors, each pixel site has its own electron to voltage conversion circuit, allowing for greater flexibility in terms of image processing and selection. However, image quality is lowered compared to CCD technology as some of the real estate within the pixel site is used for the electron to

voltage conversion circuit. Overall, CMOS technology tends to dominate the market due to its cost-effectiveness, lower power consumption, and flexibility, which translates to faster data processing.

- The *camera ECU* receives the converted voltage signals from the camera module and uses sophisticated algorithms to determine which if any ADAS subsystem, such as AEB or LKA, need to be triggered based on the current circumstances. Once the latter determination is made, the decision is transferred to the appropriate ADAS subsystem through the CAN bus. The real-time signal processing or signal control is achieved through the use of high-performance digital signal processors (DSP) and graphics processing units (GPUs).

As mentioned, cameras can be mounted at various locations on the vehicle including the front over the rearview mirror, the rear near the license plate, the sides under the sideview mirrors, the front fender, or the door pillars. For example, Subaru and Tesla use various front-facing cameras for increased object recognition accuracy. Furthermore, cabin cameras have also become popularized among auto manufacturers as a form of driver monitoring in conjunction with torque or capacitive sensors on the steering wheel.

### 7.4.2 Light Detection and Ranging Sensors

Light detection and ranging (LiDAR) is a remote sensing technology using laser light to create a 3D mapping of the surrounding environment. It is LiDAR's ability to create high-resolution 3D images of the surrounding environment at distances ranging from 100m to 200m, which makes it of great interest for ADAS applications. Additionally, LiDAR works in any lighting condition, providing it with an advantage over cameras. As a result, LiDAR and cameras are often used in conjunction to better determine distances, sizes, and speeds of objects in the vicinity of the vehicle under any lighting or weather condition. Moreover, due to the small wavelengths used in LiDAR applications, it is capable of resolving relatively smaller features at longer distances as compared to radar.

Within automotive applications, LiDAR sensors can be classified within two main categories based on their mode of laser beam steering [19, 20]:

- *Electromechanical LiDAR* is considered the first-generation LiDAR system for automotive applications. These systems use various components, including lasers and mirrors, which mechanically rotate while emitting a laser and receiving the reflected light and in turn create a 3D map of the targeted region. However, the disadvantage of these systems is their propensity to wear and tear in rough terrain due to the rapidly moving mechanical components. Additionally, these systems tend to be bulky and relatively costly.
- *Solid-state LiDAR* is a growing technology within automotive applications and consists of manufacturing a LiDAR system consisting of a laser emitter, receiver, and data processing unit all within one integrated circuit (IC) chip, which requires no rotation. The advantage of this sensor technology includes that it is cost-effective, lightweight, and compact in size allowing the placement of multiple sensors on the vehicle pointing in strategic directions for optimal object recognition.

Regardless of the LiDAR sensor used, various methods of ranging can be used that will depend on the application. These include [21]:

- Time of flight (ToF) LiDAR, which employs a pulsing laser beam and uses the amount of time that it takes for the reflected light from the detected object to reach the sensor to calculate the distance to the object. The most commonly used methods using ToF are scanning LiDAR, which uses focused rotating laser beams to cover a certain terrain and Flash LiDAR, which uses a laser beam diffused over a large region.
- Frequency-modulated continuous wave (FMCW) LiDAR [22] uses a single continuous frequency swept laser beam, which is split using a beam splitter to determine the speed of an incoming or outgoing object by analyzing the interferometric interference between the two beams.

Additionally, LiDAR sensors can also be categorized between short range and long range [23]. Short-range LiDAR has a range of approximately 25m, making it useful for blind-spot monitoring, parking assist, and lane centering. Long-range LiDAR has a range of approximately 150m, making it useful for ADAS features such as adaptive cruise control.

### 7.4.3 Radar Sensors

Radar (radio detection and ranging) follows the same working principle as LiDAR described in the previous section, but instead of light uses radio waves for its operation. Succinctly, a radar system emits radio waves in the frequency range of 24 GHz to 81 GHz [24], which then reflect off objects that are present in the vicinity of the vehicle. The reflected radio wave information is taken in by the radar sensor receiver, analyzed and used to determine the location, speed, and direction of detected objects. The longer wavelengths of radio waves, as compared to light, allow them to travel larger distances without been diffracted or distorted posing an advantage over LiDAR. Like LiDAR, radar functions in inclement weather and at night or under poor lighting conditions. However, it does not provide the same level of resolution as LiDAR. Table 7.3 [25] describes the broad radar configuration categories.

### 7.4.4 Ultrasonic Sensors

An ultrasonic sensing scheme operates by transmitting short bursts of high-frequency sound waves and measuring the amount of time that it takes for the sound to travel from the transmitter to the receiver once it is reflected from an object in the near vicinity of the vehicle.

**Table 7.3**
Radar Categories

| Category | Range (m) | Relevant ADAS |
|---|---|---|
| Short range | 0.5–20 | Blind-spot detection |
| | | Parking assist |
| | | Rear and forward collision warnings |
| Medium range | 1–60 | Front and cross traffic warnings |
| | | Lane change assist |
| | | Pedestrian detection |
| Long range | 10–250 | Automatic emergency braking |
| | | Adaptive cruise control |
| | | Forward collision warning |

*From:* [25].

Like LiDAR and radar, this time measurement can be used to calculate the distance to the detected object. Ultrasonic sensing is used in ADAS applications such as parking assist and self-parking due to its accuracy at short distances, ranging from 15 cm to 5.5m [26], and low speeds. Ultrasonic sensors are advantageous over cameras at these short distances due to a camera's poor close-distance detection [27]. Ultrasonic sensors are usually mounted on the front and rear bumpers of the vehicle and multiple sensors are used for wide coverage. The sensors must be placed at set distances from each other to prevent interference between them. Additionally, these sensors continue to perform in inclement weather and at night or under poor lighting conditions.

### 7.4.5 GPS/GNSS Sensors

The Global Navigation Satellite System (GNSS) is composed of 30 Earth-orbiting satellites capable of continuously transmitting precise time and position data. As technology moves in the direction of full automation, sophisticated and accurate GNSS sensors [28] are needed for safe, reliable, and autonomous navigation. Currently, if a GNSS sensor can at least communicate with 4 of the 30 satellites available, the vehicle position can be triangulated with an accuracy of approximately 1m. However, for an autonomous vehicle, centimeter-level accuracy is needed.

To achieve higher positional accuracy, ground-based and satellite-based augmentation systems known as GBAS and SBAS, respectively, are needed. At a high level, these augmentation systems are composed of positional reference points that are then transmitted to the satellites serving as correction factors to achieve higher positional accuracy. However, GPS/GNSS systems also have disadvantages specifically relating to poor satellite connectivity or loss of signal when traveling indoors such as in a parking garage. Additional sensors, including an inertial measurement unit (IMU) [28], can be used in conjunction with a GPS/GNSS sensor for instances when the signal is lost. IMUs can detect changes in acceleration in all directions and can perform an accurate positioning of the vehicle when the GNSS signal is not available. However, IMU's accuracy drifts over long periods of time and are only meant to be used for short periods of time if high positional accuracy is required.

### 7.4.6 Sensor Summary

All the sensors described in this section work in unison, made possible by sensor fusion, to make sense of the world surrounding the vehicle and alert the driver of any incoming or imminent hazards. Also as described, these sensors work under varying conditions, ranges, and circumstances as shown in Table 7.4 [12] and Table 7.5 [29, 30] for a fuller perception of the surrounding environment.

## 7.5 LEVEL 0 SYSTEMS

### 7.5.1 FCW and AEB

FCW and AEB systems typically use forward-facing radar and cameras to try to detect, warn, and/or react to certain objects perceived to be in the path of the vehicle that could be a potential collision target. These systems scan and monitor the roadway, processing the information through a software algorithm to evaluate the likelihood of a collision in real time. Warnings are typically activated when a threshold is reached for a time-based measure of criticality called the time-to-collision (TTC). FCW systems may issue alerts when an object is detected in front of the vehicle that may pose a risk, requiring the driver to take an evasive maneuver. These warnings may be presented

**Table 7.4**
Performance of ADAS Sensors Under Various Conditions

| | Camera | LiDAR | Radar | Ultrasonic |
|---|---|---|---|---|
| **Object Detection** | ~ | ✓ | ✓ | ✓ |
| **Object Classification** | ✓ | ~ | × | × |
| **Functionality in Inclement Weather** | × | ~ | ✓ | ✓ |
| **Functionality in Poor Lighting** | × | ✓ | ✓ | ✓ |
| **Color Perception** | ✓ | × | × | × |
| **Lane Tracking/Lane Detection** | ✓ | × | × | × |
| **Object Edge Precision** | ✓ | ✓ | × | ✓ |
| **Distance Estimation** | ~ | ✓ | ✓ | ✓ |
| **Range of Visibility** | ~ | ~ | ✓ | × |

✓ Good, ~ Fair, × Poor. *From:* [12].

**Table 7.5**
ADAS Sensors and Corresponding Vehicle Mechanism

| | Camera | LiDAR | Radar | Ultrasonic |
|---|---|---|---|---|
| **Traffic Light and Stop Sign Recognition** | ✓ | × | × | × |
| **Adaptive Cruise Control** | × | × | ✓ | × |
| **Lane Departure Warning** | ✓ | × | × | × |
| **Emergency Automatic Braking** | × | ✓ | × | × |
| **Collision Avoidance** | ✓ | ✓ | ✓ | × |
| **Park Assist** | ✓ | × | × | ✓ |
| **Pedestrian Detection** | × | ✓ | × | × |
| **Cross-Traffic Alert** | × | × | ✓ | × |
| **Rear Collision Warning** | × | × | ✓ | × |
| **Surround View** | ✓ | × | × | × |
| **Blind-Spot Detection** | ✓ | ✓ | ✓ | × |

✓ Good, × Poor. *From:* [29, 30].

individually as audible, visual, or haptic warnings or in some combination. The timing of warnings presented to the driver often changes according to the criticality of the situation and, in some systems, parameters set by the driver within a certain range. In the event that the driver does not respond to a warning, or the driver's response is inadequate to avoid an impending collision, an AEB system (if present on the vehicle) can, under certain circumstances, automatically apply the brakes in order to avoid or reduce the severity of a collision.

### 7.5.2 Lane Departure Warning and Lane Keeping Assist

A lane departure warning (LDW) system can give an early warning to the driver when the vehicle is about to or has already deviated from the lane. An LDW system does not control the vehicle actively to prevent the lane departure events. The LDW mechanism works on the principles of thresholds with respect to distance and the time to lane crossing. Data from sensors in the vehicle such as the cameras are used to provide real-time measurements and positional accuracy at up to or less than a few centimeters. This data is analyzed to determine the state of the vehicle and the warning threshold and warning time set in the system by the driver. A calculation is made to determine whether the vehicle is moving out of its current driving lane. If a determination is made to this effect, the system warns the driver through a visual, acoustic, and/or tactile signal [31]. Some challenges that LDW systems encounter for proper operation include:

- Accurate detection of lane lines specifically under all weather and visibility conditions;
- Operation under heavy traffic conditions, which may impact the lane detection capability of the system;
- Reliable operation at higher speeds, which requires the system to operate in real time even at speeds above 40 to 50 mph;
- The timing and frequency of the warning given to the driver, which should allow the driver to course correct and should also not be so frequent that it causes the driver to disable the system.

LKA systems add to the functionality of the LDW system and actively steer the vehicle towards the center of the lane to prevent lane departure. The LKA system calculates the required steering wheel torque offset to avoid a lane departure. First, the system calculates the required steering wheel angle and then compares this angle with the actual steering wheel angle. The difference between these two angles is translated into an offset torque, which is superimposed on the current steering wheel angle torque produced by the EPAS system in the vehicle [32].

## 7.6 LEVEL 1 SYSTEMS

Adaptive cruise control (ACC) and lane centering systems (LCS) are two examples of Level 1 systems. ACC systems provide two modes of control: speed and distance. When activated, these systems automatically adjust a vehicle's speed in order to maintain a preset distance with a vehicle in front. This is typically done by using a laser or radar (or both) to measure the relative distance between the host vehicle and a vehicle in front. The system provides velocity control as in conventional cruise control when there is no vehicle in front of the host vehicle. If a vehicle operates in front of the host vehicle at a slower speed, the vehicle's motor drive and braking system are controlled to maintain the intervehicle gap preset by the driver. The host vehicle returns to the preset speed again when the way ahead is no longer obstructed, resulting from either the slower vehicle ahead changing its lane or the driver of the host vehicle moving to a different lane with no vehicle in front. Some control strategies discussed in the literature for ACC systems include [33]:

- Proportional-integral-derivative (PID) control that is easy to implement and tune and robust in the presence of unmeasured disturbances but generally yields nonoptimal control in real-life driving conditions.
- Model predictive control (MPC), which is capable of real-time, multiobjective optimal control and is able to use information from GIS and GPS, but requires a certain level of future trip information, which is generally not practical in real life.
- Fuzzy logic control (FLC), which deals with the nonlinear, integrated, and complex cruise control problem with simple mathematics, but which generally yields nonoptimal control in real-life driving conditions.

While ACC systems provide automatic longitudinal control to a vehicle, LCS focus on lateral vehicle control. The LCS continuously steers the vehicle to keep it in the center of the current lane. To position the vehicle laterally, the LCS needs to know where exactly in the lane the vehicle is driving, with centimeter-level accuracy, along with an accurate knowledge of the boundaries. These systems typically use one or more in-car cameras to detect and track lane boundaries and to laterally localize the car in real time. After reliably estimating the lane boundaries along with relative position and orientation of the vehicle, the required steering angle is determined for maintaining the vehicle on a target path. The target path is continuously updated, from the estimated lane boundaries, such that the vehicle stays centered in the lane. A feedback control loop, which checks the current position and orientation of the vehicle against the target path, ensures that the vehicle stays on the target path [34]. Figure 7.5 summarizes LCS process implemented by most systems on the road today.

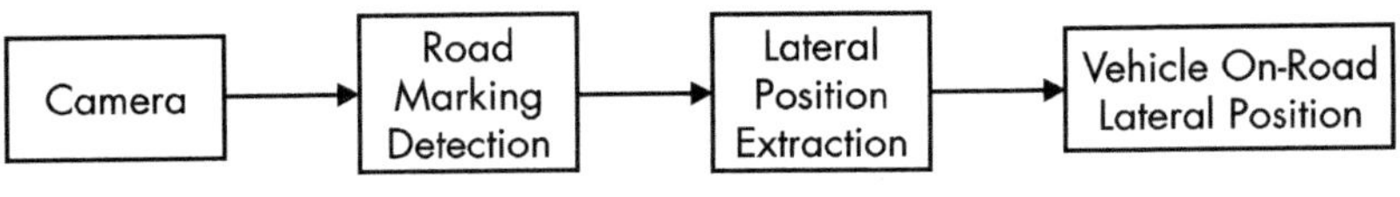


**Figure 7.5** LCS using in-vehicle cameras.

Both LCS and ACC systems are designed with the assumption that the driver determines when the engagement/disengagement of the system is appropriate. In addition, the systems are designed with the assumption that the driver continuously supervises the systems and intervenes as necessary to maintain operation of the vehicle.

## 7.7 LEVEL 2 SYSTEMS

Level 2 driving assistance systems are designed to allow the system to control both the longitudinal and lateral vehicle motion, which is done when both the LCS and ACC systems are activated at the same time. As was discussed in Section 7.6, these systems are designed with the expectation that the driver completes the object and event detection and response (OEDR)[3] subtask and supervises the driving assistance system [35]. Any Level 2 driving automation system is designed to assist and not replace the driver. All Level 2 systems require the driver to be fully responsible for driving the vehicle and to remain alert and actively supervise the system at all times. The driver has the following responsibilities when operating a Level 2 system [35]:

- Perform the remainder of the dynamic driving task not performed by the Level 2 system.
- Supervise the Level 2 system and intervene as necessary to maintain safe operation of the vehicle.
- Determine whether/when engagement and disengagement of the Level 2 system is appropriate.
- Immediately perform the entire dynamic driving task whenever required or desired.

## 7.8 LEVEL 3 SYSTEMS

Whereas a Level 2 system is considered an advanced driver support feature, a Level 3 system marks the beginning of conditional autono-

3. The subtasks of the dynamic driving task (DDT) that include monitoring the driving environment (detecting, recognizing, and classifying objects and events and preparing to respond as needed) and executing an appropriate response to such objects and events (i.e., as needed to complete the dynamic driving task and/or the dynamic driving task fallback (SAE J3016, 2016-09)).

mous driving, where drivers can legally take their eyes off the road when conditions are met. Since these systems allow the drivers to take their eyes off the road, official certifications and approvals are needed before vehicle manufacturers can put a Level 3 system on the road. In May 2022, Mercedes-Benz became the first vehicle manufacturer to be approved by the German transportation authorities to legally operate its Level 3 Drive Pilot system on the country's public roads, sold as an option on Mercedes-Benz S class and Mercedes Benz EQS [36].

## 7.9 LEVEL 4 AND LEVEL 5 SYSTEMS

Level 4 ADAS, or "high automation" technologies, are designed to perform all safety-critical driving functions and monitor roadway conditions for the entire trip. Level 4 ADAS technologies are limited to the "operational design domain" of the vehicle. In other words, the ADAS are not designed to account for every driving scenario. Presently, Level 4 ADAS are the highest level of autonomous driving technology available. Level 5 ADAS technologies, which do not require human attention and eliminate the "dynamic driving task," are still in the conceptual stage.

## 7.10 SYSTEM LIMITATIONS

The limitations of the available technology that is used for ADAS limits what these systems can do and also result in scenarios/conditions where these systems may not operate as intended. These limitations are often detailed in the owner's manuals of the vehicles incorporating these systems. This section of the chapter will provide some examples of the limitations of selected ADAS.

### 7.10.1 FCW and AEB Systems

Technological limitations can impact both the performance of FCW and AEB systems and also cause these systems to activate when there is no risk of a collision (called a false-positive activation). These technological limitations are also detailed in a National Highway Traffic Safety Administration (NHTSA) memorandum [37]:

> The AEB systems for which benefits have been documented do have limitations. Many current AEB systems activate only in front-to-rear-crash scenarios. They also may have technological limitations, such as a narrow range of operating speeds or sensors that are adversely affected by weather conditions, dirt or harsh lighting conditions. Additionally, despite extensive testing by automakers, it is possible that customers could experience unique situations resulting in unwarranted interventions by the systems.

Owner's manuals of vehicles incorporating FCW and AEB systems often provide a detailed list of conditions that can prevent these systems from operating as designed and that may also trigger false positive operation of these systems. As an example, Figure 7.6 shows the limitations of the FCW/AEB system called the Collision Mitigation braking system (CMBS) in the 2018 Honda CR-V owner's manual [38].

Table 7.6 provides a summary of some of the performance limitations of FCW/AEB systems in some 2017 through 2020 vehicles that incorporate these collision mitigation systems.

### 7.10.2 Level 2 System Limitations

The limitations of the available technology for Level 2 systems are well known and addressed in many standards including SAE J3016. As an example, SAE J3016 states that a Level 2 driving automation feature is "capable of only limited object and event detection and response, meaning that there are some events that the driving automation system is not capable of recognizing or responding to." It is for this reason that [42]: "the driver supervises the driving automation system performance by completing the object and event detection and response subtask of the dynamic driving task." It is for this reason that all vehicles equipped with Level 2 systems require the driver to always be responsible for safe operation of the vehicle due to the known technological limitations. These limitations are also often detailed in the owner's manuals of vehicles that incorporate these systems. As an example, Figure 7.7 [43] shows a section of the owner's manual for a 2019 Volvo XC60 that discusses the limitations of the vehicle's Level 2 system called Pilot Assist.

■ **With Little Chance of a Collision**

The CMBS™ may activate even when you are aware of a vehicle ahead of you, or when there is no vehicle ahead. Some examples of this are:

■ **When Passing**
Your vehicle approaches another vehicle ahead of you and you change lanes to pass.

■ **At an intersection**
Your vehicle approaches or passes another vehicle that is making a left or right turn.

■ **On a curve**
When driving through curves, your vehicle comes to a point where an oncoming vehicle is right in front of you.

■ **Through a low bridge at high speed**
You drive under a low or narrow bridge at high speed.

■ **Speed bumps, road work sites, train tracks, roadside objects, etc.**
You drive over speed bumps, steel road plates, etc., or your vehicle approaches train tracks or roadside objects [such as a traffic sign and guard rail] on a curve or, when parking, stationary vehicles and walls.

Collision Mitigation Braking System™ (CMBS™)*

For the CMBS™ to work properly:

Always keep the radar sensor cover clean.

Never use chemical solvents or polishing powder for cleaning the sensor cover. Clean it with water or a mild detergent.

Do not paint, or apply any coverings or paint to radar sensor area. This can impact CMBS™ operation.

Have your vehicle checked by a dealer if you find any unusual behavior of the system (e.g. the warning message appears too frequently).

If the front of the vehicle is impacted in any of the following situations, the radar sensor may not work properly. Have your vehicle checked by a dealer:
- The vehicle mounted onto a bump, curb, chock, embankment, etc.
- You drive the vehicle where the water is deep.
- Your vehicle has a frontal collision.

**Figure 7.6** CMBS limitations listed in 2018 CR-V owner's manual.

**Table 7.6**
FCW/AEB System Limitations in Selected 2017 to 2020 Vehicles

| | 2017 Lexus GS 450H [39] | 2020 Subaru Forester [40] | 2018 Genesis G80 [41] |
|---|---|---|---|
| **System Name** | Pre-Collision System | Pre-Collision Braking System | Automatic Emergency Braking |
| **Sensors Used** | Radar and Camera | Radar and Camera | Radar and Camera |
| **Conditions Under Which System May Operate When There Is No Possibility of a Collision** | When overtaking a preceding vehicle that is changing lanes<br>When driving on a road where relative location to vehicle ahead in an adjacent lane may change such as on a winding road<br>When approaching objects on the roadside, such as guardrails, utility poles, trees, or walls<br>When driving on a narrow path surrounded by a structure, such as in a tunnel or on an iron bridge<br>When there is a metal object (manhole cover, steel plate), steps, or a protrusion on the road surface or roadside<br>When passing through a place with a low structure above the road (low ceiling, traffic sign) | Passing hanging banners, flags, or branches<br>Passing through an automatic gauge<br>Driving in a location where the grade of the road changes rapidly<br>There is an obstacle on a curve or intersection<br>A vehicle or an object is being narrowly passed<br>Stopping very close to a wall or a vehicle in front | The vehicle is driven near areas containing metal substances such as a construction zone, railroad<br>In certain instances on a curved road<br>When driving on a curved road with a vehicle in the adjacent lane<br>When driving upwards or downwards on a slope |

This table does not provide an exhaustive list of all conditions listed in the vehicle's owner manuals.

As another example, Figure 7.8 [44] shows a section of the owner's manual of a 2018 Lexus GS 450, which discusses the limitations of the Level 0 lane keeping assist system incorporated into the vehicle.

Table 7.7 provides a summary of Level 2 systems in some 2017 model year vehicles and some of the performance limitations of these systems.

- Pilot Assist is not a substitute for the driver's attention and judgment. The driver is always responsible for ensuring the vehicle is driven in a safe manner, at the proper position within the lane, at the appropriate speed, with an appropriate distance to other vehicles, and in accordance with current traffic rules and regulations.

**Figure 7.7** Section of the Pilot Assist-related warnings in the 2019 XC60 Owner's Manual. (*From:* [43].)

⚠ WARNING

■ **Before using LKA system**

Do not rely solely upon the LKA system. The LKA system does not automatically drive the vehicle or reduce the amount of attention that must be paid to the area in front of the vehicle. The driver must always assume full responsibility for driving safely by paying careful attention to the surrounding conditions and operating the steering wheel to correct the path of the vehicle. Also, the driver must take adequate breaks when fatigued, such as from driving for a long period of time.
Failure to perform appropriate driving operations and pay careful attention may lead to an accident, resulting in death or serious injury.

**Figure 7.8** Lane Keeping Assist warning in the 2018 Lexus GS 450h owner's manual. (*From:* [44].)

## 7.11 FUNCTIONAL SAFETY AND ADAS

The concept of functional safety was discussed in Chapter 3 in the context of EV battery BMS. Functional safety-related issues for some ADAS are comparatively more complex as compared to other vehicle electronic control systems. This is because the function performed by these ADAS makes it necessary to incorporate "fail operational" properties which requires the system to be designed in a manner that if a ADAS component were to fail, the system is designed with safeguards in place to allow for the safe continuation and eventual termination of the vehicle's path without resulting in a hazard. Additional com-

**Table 7.7**
Performance Limitations of Level 2 Systems in Some 2017 Vehicles

| | **2017 Lexus GS 450h [39]** | **2017 Mercedes S450 [45]** | **2017 Volvo XC90 [46]** |
|---|---|---|---|
| **Level 2 System Name** | Lane Keep Assist (Lane Centering Function) | DISTRONIC PLUS with Steering Assist and Stop&Go Pilot | Pilot Assist |
| **Function** | Provides the required assistance by operating the steering wheel to keep the vehicle in its current lane | Helps keep the vehicle in the center of the lane by means of moderate steering interventions in a speed range from 0–125 mph | Helps keep the vehicle in its current traffic lane by providing steering assistance and maintaining an even speed and a set time interval to the vehicle ahead |
| **Operating Speed Range (MPH)** | >32 | <125 | <87 |
| **Conditions That May Lead to Performance Impact** | Light from the headlights of an oncoming vehicle, the sun enters the camera<br>The vehicle is driven where the road diverges, merges<br>The traffic lane is excessively narrow or wide<br>The white (yellow) lines are cracked, "Botts' dots", "Raised pavement market" or stones are present | Poor visibility (e.g., insufficient road illumination, snow, rain, fog)<br>If there is glare (e.g., from oncoming traffic, the Sun, or reflection from other vehicles)<br>There are no, several, or unclear lane markings for a lane<br>The lane markings are worn away, dark, or covered up<br>The lane markings change quickly (e.g., lanes branch off, cross one another, or merge) | Lane markings are missing, badly faded, or cross each other<br>The division of lanes is not clear (e.g., when a lane divides or merges with another, at exits, or if there are many road signs/markers)<br>Bad weather conditions (rain, snow, fog, slush, poor visibility, back lighting) |

This table does not provide an exhaustive list of all conditions listed in the vehicle's owner manuals.

plexity involves the need to design the system with an assurance of a reasonable and predictable life span.

Level 0 through Level 3 ADAS require the driver to always be driving the vehicle and supervising the operation of the ADAS. As such, when a failure is detected in one of these systems, the human driver can be asked to take control of the vehicle to bring the vehicle to a safe stop. This means that the functional safety concepts applicable to the development of these systems are not very different from the functional safety concepts already discussed in Chapter 3 in the context of an EV battery BMS. These include [47]:

- Carrying out a safety analysis in each phase of the system's development process to identify failures/conditions that may lead to a hazardous condition or violate a safety requirement that was already derived in a previous phase of the development cycle.
- To use a combination of severity classes and ASIL ratings to develop appropriate safety goals.
- To identify root causes that can lead to failures that may violate the safety goals.
- To build safety requirements that address what the system must perform to detect a fault and what actions the system must perform on fault detection so that the subsequent hazard can be avoided.

For Level 4 and Level 5 ADAS, the safety analysis is comparatively more difficult than that for Level 0 through Level 3 systems because the driver is not required to be in control of the vehicle, which means that any malfunction must be handled by another function in the vehicle. The limits between these functions become blurred and difficult to define because the interaction between the different functions is now more complex. The concept phase of ISO 26262 becomes more important for these functions because the development of these systems requires the engineering approaches and technologies beyond state of the art including the influence of the driver in the hazard and risk assessment (HARA) phase, the definition of safety goals and corresponding attributes for specific levels of the system (e.g., safe state) and the changes of the functional safety concept from fail-safe to fail-operational strategies [48].

## 7.12 SUMMARY

The number of vehicles on the road equipped with ADAS is increasing exponentially. Agencies such as the NHTSA are starting to mandate certain ADAS features in vehicles including making AEB, including pedestrian AEB, standard on all cars and light trucks by September

2029 [49]. As sensor and computing processing technology continues to improve, it is anticipated that even more advanced ADAS features will start to be incorporated in vehicles. While Level 4 through Level 5 systems are not available for consumers to buy and experience on their own, continuous technological advancements will likely make that a reality in the not-too-distant future.

## References

[1] Martin, H., et al., "Functional Safety of Automated Driving Systems: Does ISO 26262 Meet the Challenges?" September 2017, DOI:10.1007/978-3-319-31895-0_16.

[2] SAE J3016, "Taxonomy & Definitions for Terms Related to Driving Automation Systems for On-Road Motor Vehicles," April 2021.

[3] https://cars.usnews.com/cars-trucks/features/mercedes-benz-introduces-level-3-autonomy-will-it-work.

[4] https://www.synopsys.com/automotive/autonomous-driving-levels.html.

[5] NHTSA, "Driver Assistance Technologies," https://www.nhtsa.gov/equipment/driver-assistance-technologies.

[6] Partnership for Analytics Research in Traffic Safety (PARTS), "Real-World Effectiveness of Model Year 2015-2020 Advanced Driver Assistance Systems," 2022.

[7] IIHS-HLDI, "Real-World Benefits of Crash Avoidance Technologies," 2022.

[8] Cicchino, J. B., "Effects of Automatic Emergency Braking Systems on Pedestrian Crash Risk," Insurance Institute for Highway Safety, 2022.

[9] IIHS-HLDI, "Petition for Rulemaking," March 2022.

[10] NHTSA, "Automated Vehicles for Safety," https://www.nhtsa.gov/technology-innovation/automated-vehicles-safety.

[11] NHTSA, "The Economic and Societal Impact of Motor Vehicle Crashes," 2019.

[12] Kumar Sarkar, P., "Driver Assistance Technologies," IntechOpen, 2020, DOI: 10.5772/intechopen.94345.

[13] De Jong, Y., et al., "Sensor and Sensor Fusion Technology in Autonomous Vehicles: A Review," *Sensors*, Vol. 21, 2021, p. 2140.

[14] LeddarTech Inc., "The Need, Techniques and Advantages of Raw Data Sensor Fusion in ADAS and AD for Sensing and Perception," Technical Note, 2023.

[15] Krose, B. J. A., et al., "An Introduction to Neural Networks," *Journal of Computer Science*, January 1993.

[16] https://towardsdatascience.com/the-basics-of-neural-networks-neural-network-series-part-1-4419e343b2b.

[17] https://www.hitachi-automotive.us/Products/oem/DCS/ADAS/index.htm.

[18] https://www.rohm.com/solution/automotive/adas/adas-camera-system.

[19] https://www.eetimes.com/the-future-of-lidar-lies-in-adas/.

[20] https://www.einfochips.com/blog/how-lidar-based-adas-work-for-autonomous-vehicles/.

[21] https://www.foresightauto.com/an-overview-of-autonomous-sensors-lidar-radar-and-cameras/.

[22] https://www.bridgerphotonics.com/blog/frequency-modulated-continuous-wave-fmcw-lidar.

[23] https://www.hesaitech.com/essential-guide-to-adas-automation-systems-involving-lidar/.

[24] https://cetecomadvanced.com/en/news/frequency-ranges-for-automotive-radar-technology/.

[25] https://caradas.com/adas-radar-sensor/.

[26] https://www.bosch-mobility.com/en/solutions/sensors/ultrasonic-sensor.

[27] https://www.electronicdesign.com/markets/automotive/article/21805470/radar-and-ultrasonic-sensors-strengthen-adas-object-detection.

[28] https://dewesoft.com/blog/types-of-adas-sensors#gps-gnss-sensors.

[29] https://e2e.ti.com/blogs_/b/behind_the_wheel/posts/cars-are-becoming-rolling-sensor-platforms.

[30] https://www.futurebridge.com/industry/perspectives-mobility/gated-sensor-in-adas-improved-vehicle-safety/.

[31] Chen, W., et al., "Lane Departure Warning Systems and Lane Line Detection Methods Based on Image Processing and Semantic Segmentation: A Review," *Journal of Traffic and Transportation Engineering*, 2020.

[32] Pohl, J., et al., "A Driver-Distraction-Based Lane-Keeping Assistance System," *Systems and Control Engineering*, IMechE Vol. 221, 2007.

[33] He, Y., et al., "Adaptive Cruise Control Strategies Implemented on Experimental Vehicles: A Review," *IFAC PapersOnLine*, Vol. 52, No. 5, 2019, pp. 21–27.

[34] Rameez, I., "Next-Generation Lane Centering Assist System," University of Technology, Eindhoven, 2017.

[35] SAE J3016, "Taxonomy and Definitions for Terms Related to Driving Automation Systems for On-Road Motor Vehicles," 2016-09, p. 23.

[36] https://www.motortrend.com/news/mercedes-benz-drive-pilot-eqs-autonomous-driverless-first-drive-review/.

[37] US DOT/NHTSA, "Docket Submission of Commitments to Advancing Automatic Emergency Braking Technology," Docket Number NHTSA-2015-0101, March 25, 2016.

[38] https://techinfo.honda.com/rjanisis/pubs/OM/AH/ATLA1818OM/enu/ATLA1818OM.PDF.

[39] 2017 Lexus GS 450h Owner's Manual.

[40] 2020 Subaru Forester EyeSight Manual.

[41] 2018 Genesis G80 Owner's Manual.

[42] SAE J3016, "Taxonomy and Definitions for Terms Related to Driving Automation Systems for On-Road Motor Vehicles," 2018-06, p. 24.

[43] https://volvornt.harte-hanks.com/manuals/2019/XC60_OwnersManual_MY19_en-US_TP25629.pdf.

[44] https://assets.sia.toyota.com/publications/en/om-s/OM30G62U/pdf/OM30G62U.pdf.

[45] S-Class Operator's Manual, 2017.

[46] 2017 XC90 Owner's Manual.

[47] Martin, K., et al., "Automated Functional Safety Analysis of Automated Driving Systems, Formal Methods for Industrial Critical Systems," *23rd International Conference*, Ireland, 2018.

[48] Martin, H., K. Tschabuschnig, O. Bridal, and D. Watzenig, "Functional Safety of Automated Driving Systems: Does ISO 26262 Meet the Challenges?" in *Automated Driving*, D. Watzenig, and M. Horn (eds), Springer, Cham, 2017.

[49] https://www.nhtsa.gov/press-releases/nhtsa-fmvss-127-automatic-emergency-braking-reduce-crashes.

# 8

# BATTERY-RELATED STANDARDS AND REGULATIONS

A number of standards have been developed around the world to target battery systems in EVs. Standards are often voluntary documents drafted by nongovernmental organizations such as the Society of Automotive Engineers (SAE), the International Electrotechnical Commission (IEC), and Underwriters Laboratories (UL). Regulations are issued by governmental authorities and have the force of law. For vehicles in the United States, the National Highway Traffic Safety Administration (NHTSA) issues regulations via the Federal Motor Vehicle Safety Standards (FMVSS), which sets minimum safety performance requirements for motor vehicles or motor vehicle equipment. In Europe, type approval regulations are issued by the United Nations Economic Commission for Europe (UNECE). These regulations define uniform technical prescriptions for vehicles and their parts and equipment. In 2012, EU and U.S. standards organizations agreed to a Transatlantic Cooperation on standards for EVs to avoid the development of conflicting EV and battery safety standards [1].

Table 8.1 lists some of the EV-related standards and regulations in Europe, Canada, and the United States. In Europe, batteries, regardless of their type or application, are covered under the EU Battery Directive (2006/66/EC and Amendment 2013/56/EU). This directive restricts the use of lead to 0.004%, mercury to 0.0005%, and cadmium to 0.002%. In addition, it is common for protection electronics

**Table 8.1**
EV-Related Standards and Regulations in Europe and North America

| Standard | Description | Applicable Region | Applicability Level |
|---|---|---|---|
| ECCE R100 | Safety requirement with respect to the electric power train of road vehicles | Europe | Pack |
| FMVSS 305 | Contains specific requirements for limitation of electrolyte spillage and retention of electric energy storage/ conversion devices during and after a crash | United States | Pack and module |
| CMVSS 305 | Canadian equivalent of FMVSS 305 | Canada | Pack and module |
| UN 38.3 | UN Transportation requirements for lithium batteries | Global | Pack, module, and cell |
| GHS | UN Globally Harmonized System of Classification and Labelling of Chemicals | Global | Pack, module, and cell |
| DGR | International Air Transport Association (IATA) - Dangerous Goods Regulations | Global | Pack, module, and cell |
| IMDGR | International Maritime Dangerous Goods Regulation | Global | Pack, module, and cell |
| RID | OTIF-Regulation concerning the International Carriage of Dangerous Goods by Rail | Europe | Pack, module, and cell |
| ADR | European Agreement concerning the International Carriage of Dangerous Goods by Road | Europe | Pack, module, and cell |
| Directive 2006/66/EC | Batteries Directive (Hazardous substance restriction requirement) | Europe | Pack, module, and cell |

used in EV battery modules and battery packs to comply with the Restriction of Hazardous Substances (RoHS) directive.

## 8.1 PERFORMANCE TESTS

Several standards have been developed to provide a methodology for determining the performance of Li-ion cells, modules, and battery packs in EVs. One commonly used standard that provides test procedures and conditions for testing the basic performance characteristics of Li-ion batteries including capacity, power density, energy density, storage life, and cycle life is IEC 62660-1 (Secondary Lithium-ion Cells for the Propulsion of Electric Road Vehicles, Part 1: Performance Testing). Table 8.2 provides a high-level overview of the cell performance tests in IEC 62660-1.

**Table 8.2**
IEC 62660-1 Performance Tests for Li-Ion Cells

| Test Item | Test Type | Condition | Requirement |
|---|---|---|---|
| Energy & Capacity | Constant Discharge | Temperature (°C) | −20, 0, 25, 45 |
| | | Charge | Standard charge |
| | | Discharge | 1C, 10C, 20C, Imax, C/3, 2C, and 5C |
| | | Cycles | 2 |
| Power & Resistance | Charge/discharge pulse test | Temperature (°C) | −20, 0, 25, 45 |
| | | Discharge | 0.2C, 1C, 5C, 10C, C/3, 2C, Imax |
| | | Charge | 1/3C, 1C, 2C, 5C, 10C, Imax |
| | | Duration | 10 seconds |
| | | SOC | 50% |
| Energy Efficiency | Pulse charge/discharge | Temperature (°C) | −20, 0, 25, 45 |
| | | Discharge | 1/3C, 1C, 5C, 10C |
| | | Charge | 1/3C, 1C, 5C, 10C |
| | | Duration | 10 seconds |
| | | SOC | 100% |
| Storage Test | Aging and lifetime of a cell under storage or nonuse | Storage | Constant temperature and constant voltage |
| | | Periodic checkup | Performed every 42 days |
| Cycle Life Test | Charge/discharge cycling | Cycle parameters | Application-dependent |

## 8.2 SAFETY TESTS

Li-ion batteries used in EVs must pass a series of safety tests to be certified for use in an application. Safety tests are described in international, national, and regional standards that are typically developed based on prenormative research and experience from industry, academia, and regulatory bodies. These tests are often performed to understand and identify potential weak points in a battery design and to understand how a battery will behave under abuse conditions. Table 8.3 provides a list of commonly used safety standards for EV batteries in Europe and North America.

Abuse tests in the various safety standards are often classified by the nature of the abuse: mechanical, electrical, environmental, and chemical. Most standards set "no fire" and "no explosion" as the pass criteria for the tests. EUCAR hazard levels are also sometimes

**Table 8.3**
Safety Standards for EV Batteries in Europe and North America

| Standard | Description of This Standard | Applicable Region | Applicability Level |
|---|---|---|---|
| ISO 6469-1 | This document specifies safety requirements for rechargeable energy storage systems (RESS) of electrically propelled road vehicles for the protection of persons | Global | Pack |
| ISO 12405-4 | This document specifies test procedures for the basic characteristics of performance, reliability, and electrical functionality for the battery packs and systems for either high-power or high-energy application | Global | Pack |
| IEC 62660-2 | This document specifies test procedures to observe the reliability and abuse behavior of secondary Li-ion cells and cell blocks used for the propulsion of EVs | Global | Cell |
| ISO 26262 | This document specifies functional safety of electrical and electronic systems | Global | Pack and module |
| SAE J2344 | Technical guidelines relating to safety for vehicles that contain high voltage | North America | Pack |
| SAE J2464 | This document describes a body of tests that may be used as needed for abuse testing of EVs or hybrid EV RESS to determine the response of such electrical energy storage and control systems to conditions or events that are beyond their normal operating range | North America | Pack, module, and cell |
| SAE J2929 | This SAE standard defines a minimum set of acceptable safety criteria for a lithium-based rechargeable battery system to be considered for use in a vehicle propulsion application as an energy storage system connected to a high-voltage power train | North America | Pack |
| UL 2580 | This standard evaluates the electrical energy storage assembly's ability to safely withstand simulated abuse conditions and prevents any exposure of persons to hazards as a result of the abuse | North America | Pack and module |
| Freedom CAR SAND 2005-3123 | This manual defines a complete body of abuse tests intended to simulate actual use and abuse conditions that may be beyond the normal safe operating limits experienced by electrical energy storage systems used in electric and hybrid EVs | North America | Pack, module, and cell |

used to classify the response of the tested module or battery to a particular abuse condition. EUCAR hazard levels range from Level 0 (no effect, system maintains its functionality) to Level 7 (explosion, disintegration of the cell is observed) and are defined as follows:

- Level 0 (No Effect): No loss of functionality.
- Level 1 (Passive Protection Activated): Cell reversibly damaged and a repair of a protection device is needed.
- Level 2 (Defect Damage): Cell irreversibly damaged.
- Level 3 (Leakage, mass change < 50%): Weight loss of <50% of electrolyte weight but no fire, flame, explosion, or rupture.
- Level 4 (Venting, mass change > 50%): Weight loss of >50% of electrolyte weight but no fire, flame, explosion, or rupture.
- Level 5 (Fire or Flame): No rupture, explosion, or flying parts.
- Level 6 (Rupture): No explosion but flying parts of the active mass observed.
- Level 7 (Explosion): Explosion and disintegration of the cell observed.

### 8.2.1 Mechanical Abuse Tests

Mechanical abuse tests can be performed at the cell level, the module level, the battery pack level, or even the vehicle level. These tests simulate a variety of different abuse conditions and characterize the response of the vehicle structure and the battery pack to the abuse condition to understand how the vehicle and its battery may react to a crash on the road. Table 8.4 provides a list of the mechanical abuse tests in the different standards that are often used to evaluate an EV battery. The different standards provide protocols for running these tests at the cell level, the module level, or the battery pack level.

#### *8.2.1.1 Vibration Test*

Most abuse standards require the battery packs to pass a vibration test that aims to evaluate the effect of long-term vibration profiles that are representative of a vehicle driving with the battery pack to assess durability. However, many of the vibration profiles described in the standards represent short-term abuse rather than a more longer-term durability test that would better represent battery life. The vibration profiles in the standards are either sine waves or random profiles and are derived from measurements from conventional vehicles at loca-

**Table 8.4**
Mechanical Abuse Tests in Different Standards
(C: Cell Level; M: Module Level; P: Pack Level)

| Test | UN 38.3 | UL 2580 | ECE R100 | SAE J2464 | SAE J2929 | ISO 6469-1 | IEC 62660-2 | FreedomCAR |
|---|---|---|---|---|---|---|---|---|
| Vibration | C, M, P | M, P | P | | P | P | C | |
| Mechanical Shock | C, M, P | M, P | P | C or above | P | P | C | M |
| Impact | C | | | | | | | |
| Drop | | M, P | | P | P | | | P |
| Penetration | | | | C or above | | | | C |
| Rollover | | M, P | | M, P | P | | | M |
| Crush (Mechanical Integrity) | | M, P | P | C or above | P | | C | M, P |

tions typically used to mount battery packs in EVs. Table 8.5 lists the frequency range and the peak acceleration for the vibration test at the battery pack level in the different standards.

**Table 8.5**
Test Frequency Range and Peak Acceleration for the Vibration Test at the Battery Pack Level in the Different Standards

| Standard | Frequency Range (Hz) | Peak Acceleration (g) |
|---|---|---|
| UN 38.3 | 7–200 | 1 |
| UL 2580 | 10–190 | 1.1 |
| ECE R100 | 7–50 | 1 |
| SAE J2929 | 10–190 | 1.1 |
| ISO 6469-1 | 5–190 | |
| IEC 62660-2 | 10–2,000 | 2.0 |
| FreedomCAR | 10–190 | 2.5 |

#### 8.2.1.2 Mechanical Shock Test

The aim of the mechanical shock test is to characterize the response of the battery to a sudden acceleration or deceleration condition that may occur while the battery is installed in the vehicle. During the mechanical shock test, the tested sample is exposed to a shock force defined in terms of acceleration and a shock duration adapted to different conditions from normal driving to a vehicle crash condition. The actual test conditions in the different standards show many differences in terms of acceleration and shock duration. Some standards (e.g., SAE J2464 and FreedomCAR) also divide the shock test into different levels/grades. While most standards require a half-sine wave pulse, some standards allow the use of other pulse shapes to simulate actual deceleration pulses more accurately. Some standards (e.g., ECE R100) also provide different peak acceleration requirements for different vehicle masses. Table 8.6 shows a comparison of peak accelerations and shock durations for testing battery packs in the different standards.

#### 8.2.1.3 Drop Test

The goal of the drop test is to assess the response of a battery that is being installed or removed from an EV to being accidentally dropped. Some standards (e.g., SAE J2929) require the test to be performed at a height that is determined to be the maximum distance to ground or

**Table 8.6**
Comparison of Peak Accelerations and Shock Duration for Testing Battery Packs in the Different Standards

| Standard | Peak Acceleration (g) | Maximum Duration (ms) |
|---|---|---|
| UN 38.3 | 50 | 11 |
| UL 2580 | 25 | 15 |
| ECE R100 | 28 | 80 |
| SAE J2464 | 25 | 60 |
| SAE J2929 | 25 | 60 |
| ISO 6469-1 | 7 | 6 |
| IEC 62660-2 | 25 | 15 |
| FreedomCAR | 25 | 60 |

maximum possible drop distance that the battery system may experience when serviced according to documented procedures. Table 8.7 summarizes the drop test conditions for standards that contain requirements for performing drop tests at the battery pack level.

#### 8.2.1.4 *Rollover Test*

Also often referred to as the rotation test, this test simulates the rollover of a vehicle during an accident. Regulations (e.g., FMVSS 305) require the rollover test to be performed after the vehicle goes through a crash test. The test involves the slow rotation of a battery pack to evaluate the development of electrolyte or coolant leakage from the battery pack. All standards have a very similar rotation speed requirement of 90°/15 seconds for the test at the battery pack level. The different standards require different monitoring intervals after the rotation is completed.

#### 8.2.1.5 *Crush Test*

The goal of the crush test in the various standards is to simulate the potential deformation/damage of the battery enclosure as a result of a vehicle accident. Most standards require the crushing to be performed using a ribbed test platen such as the platen surface shown in Figure 8.1 [2]. Table 8.8 provides a summary of the required crush force when testing battery packs in the different standards.

### 8.2.2 Electrical Abuse Tests

Just like the mechanical abuse tests, the electrical abuse tests can also be performed at the cell level, the module level, or the battery pack level. These tests simulate the response of the battery pack to different electrical abuse conditions that may result in the cells operat-

**Table 8.7**
Pack-Level Drop Test Conditions in Different Standards

| Standard | Drop Height (m) | Number of Drops | Surface |
|---|---|---|---|
| UL 2580 | 1 | 1 | Flat concrete |
| SAE J2464 | 10 | 1 | Cylindrical steel object |
| SAE J2929 | >1 | 1 | Flat surface |
| FreedomCAR | Up to 10 | 1 | Cylindrical steel object |

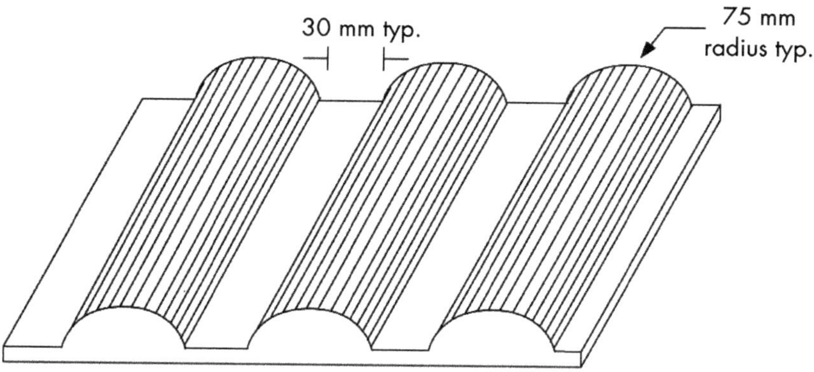


**Figure 8.1** Platen surface for the crush test in the FreedomCAR standard.

**Table 8.8**
The Required Crush Force When Testing Battery Packs in the Different Standards

| Standard | Crush Force |
|---|---|
| UL 2580 | 100 ± 6 kN |
| ECE R100 | 100 kN |
| SAE J2464 | 1,000 times battery weight |
| SAE J2929 | 100 kN |
| FreedomCAR | 1,000 times battery weight |

ing outside their rated specifications. Table 8.9 provides a list of the electrical abuse tests in the different standards that are often used to evaluate EV battery packs.

### *8.2.2.1 External Short-Circuit Test*

The goal of this test is to characterize the response of the over-current protection circuit/device during an external short-circuit condition to determine if the battery pack can withstand a cell over-current condition that can cause one or more cells to go into the thermal runaway and to determine if the high-voltage bus bars and other infrastructure in the battery pack can handle the short-circuit condition without resulting in damage to circuitry, arcing, or a loss or reduction in isola-

**Table 8.9**
Electrical Abuse Tests in Different Standards
(C: Cell Level; M: Module Level; P: Pack Level)

| Test | UN 38.3 | UL 2580 | ECE R100 | SAE J2464 | SAE J2929 | ISO 6469-1 | IEC 62660-2 | FreedomCAR |
|---|---|---|---|---|---|---|---|---|
| External Short-Circuit | C, M, P | M, P | P | C, M, or P | P | P | C | M |
| Overcharge | C, M, P | M, P | P | C, M, or P | | P | C | M |
| Forced Discharge (Over-discharge) | C | M, P | P | C, M | | P | C | M |
| Imbalanced Charging Test | | M, P | | | | | | |
| Dielectric Voltage Withstand Test | | M, P | | | | | | |
| Isolation Resistance Test | | M, P | | | | | | |
| Continuity Test | | M, P | | | | | | |

tion resistance.[1] All standards require this test to be performed on a fully charged battery with some standards (e.g., SAE J2464) requiring the bypassing of any nonpassive short-circuit protection devices before performing the test. Table 8.10 lists the external short-circuit resistance and battery temperature requirements for the external short-circuit test in the different standards.

### *8.2.2.2 Overcharge/Over-Discharge Test*

The overcharge/over-discharge test is performed to evaluate the ability of the battery protection to prevent the cells from operating outside their rated voltages under foreseeable failure conditions in the field (e.g., charger failure). A number of standards (e.g., SAE J2464) also require nonpassive protection devices to be disabled during the test.

1. Isolation resistance in a battery pack refers to the measure of electrical insulation in the battery pack and provides an indication of the resistance to current flow through unintended paths (e.g., insulation materials surround the cells) in the battery pack.

**Table 8.10**
External Short-Circuit Resistance and Battery Temperature Requirements for the External Short-Circuit Test in Different Standards

| Standard | Resistance (mΩ) | Battery Temperature for Test (°C) |
|---|---|---|
| UN 38.3 | <100 | 57 ± 4 |
| UL 2580 | <20 | 25 ± 5 |
| ECE R100 | <5 | 20 ± 10 |
| SAE J2464 | <5 | 25 ± 3 |
| SAE J2929 | <5 | 25 ± 3 |
| ISO 6469-1 | <5 | 25 ± 3 |
| IEC 62660-2 | <5 | 20 ± 10 |
| FreedomCAR | <5 | Nominal operating temperature |

Table 8.11 summarizes the overcharge test protocols in the different standards.

#### *8.2.2.3 Other Electrical Abuse Tests*

UL 2580 is one of the standards that contains a number of other electrical abuse tests for EV battery packs. These include:

**Table 8.11**
Overcharge Test Protocols in the Different Standards

| Standard | Charge Current | End Condition (If Protection Circuit Does Not Operate) | Faults Simulated |
|---|---|---|---|
| UN 38.3 | Twice manufacturer's recommended maximum continuous charge current | 1.2 times maximum charge voltage or 24 hours | None |
| UL 2580 | Maximum manufacturer specified charge rate | 200% SOC or 110% rated capacity | Single fault in charger circuit |
| ECE R100 | Maximum manufacturer specified charge rate | 10°C above maximum operating temperature specified by manufacturer | None |
| SAE J2464 | 32A | 200% SOC or 4 hours | Nonpassive protection devices |
| SAE J2929 | 1C | 200% SOC | None |
| ISO 6469-1 | 1C | 200% SOC | None |
| IEC 62660-2 | 1C | 200% SOC | None |
| FreedomCAR | 32A | 200% SOC or 4 hours | None |

- Imbalanced charging test, which is performed to determine if the battery pack with series-connected modules can maintain the cells/modules within their specified operating parameters if the modules become imbalanced.
- Dielectric voltage withstand test, which evaluates the electrical spacings and insulation of any hazardous voltage circuits in the battery pack.
- Isolation resistance test, which evaluates whether the insulation in the battery pack provides adequate isolation of hazardous voltage circuits from accessible conductive paths.
- Continuity test, which evaluates the continuity of any protective grounding circuit in the battery pack.

### 8.2.3 Thermal/Environmental Tests

The thermal/environmental tests evaluate the safety performance of the battery system under conditions of temperature changes such as due to a vehicle fire after a crash or extreme weather conditions in some parts of the world. Table 8.12 provides a list of the thermal/environmental tests in the different standards that are often used to evaluate EV battery packs.

#### *8.2.3.1 Thermal Cycling/Shock Test*

The aim of the thermal cycling/shock test is to characterize the impact of any changes due to the expansion and contraction of battery pack components upon exposure to sudden extreme changes in temperature. The standards have different requirements on the extreme temperatures that should be tested, the time at each temperature extreme, and even the transition time between the temperature extremes (Table 8.13).

#### *8.2.3.2 Fire Exposure Test*

The fire exposure test is performed on the battery pack to assess the risk of an explosion if the EV or its battery pack is exposed to an external fire. The different standards have given different names to this test and have their own unique protocols for conducting this test. As an example:

**Table 8.12**
Thermal/Environmental Tests in Different Standards
(C: Cell Level; M: Module Level; P: Pack Level)

| Test | UN 38.3 | UL 2580 | ECE R100 | SAE J2464 | SAE J2929 | ISO 6469-1 | IEC 62660-2 | FreedomCAR |
|---|---|---|---|---|---|---|---|---|
| Thermal cycling/ shock test | C, M, P | M, P | P | C, M, P | P | P | C | C |
| High-temperature hazard test (fire exposure test) | | M, P | P | M, P | P | P | | M |
| Thermal stability test | | | | C, M, P | | | C | C, M, P |
| Cycling without thermal management | | P | P | M, P | P | | | |
| Altitude simulation | C, M, P | | | | | | | |
| Immersion test | | M, P | | M or P | P | P | | M, P |

**Table 8.13**
Comparison of the Thermal Cycling/Shock Test Requirements in the Different Standards

| Standard | Temperature Extremes (°C) | Time at Each Temperature Extreme (hours) | Transition Time Between Temperature Extremes (minutes) | Number of Cycles |
|---|---|---|---|---|
| UN 38.3 | –40 to 72 | 12 | 30 | 10 |
| UL 2580 | –40 to 85 | >6 | 15 | 5 |
| ECE R100 | –40 to 60 | 6 | 30 | 5 |
| SAE J2464 | –40 to 80 | >6 | 15 | 5 |
| SAE J2929 | –40 to 70 | 6 | 15 | 5 |
| IEC 62660-2 | –20 to 65 | 6 | 30 | 30 |
| FreedomCAR | –40 to 80 | >6 | 30 | 5 |

- SAE J2464 requires the performance of the radiant heat test, which requires the battery pack to be placed in a radiant heating fixture that is programmed to reproduce the temperature experienced in a fuel fire (890°C). The programmed temperature is achieved within 90 seconds and held for 10 minutes.

- UL 2580 requires the battery packs to pass an external fire exposure test, which exposes the battery pack to a uniform fire source along its length at its bottom surface. The test setup is required to achieve a temperature of 590°C within 5 minutes of ignition (as measured through thermocouples instrumented on the battery pack for the test).
- ECE R100 requires the battery packs to pass a fire resistance test where the battery pack is exposed to a fire obtained by burning commercial fuel in a pan.

#### 8.2.3.3 *Thermal Stability Test*

The goal of the thermal stability test is to identify the temperature at which the battery pack goes into thermal runaway. This test can be performed at the cell, the module, or the battery pack level and typically involves the sequential increase of temperature in fixed temperature steps with a holding time at each incremental step. As an example, SAE J2424 requires the battery pack to be placed in a device or chamber capable of heating the battery pack to 200°C above its operating temperature (if the temperature at which the battery pack may go into thermal runaway is not known). The temperature is increased in increments of 20°C and then held at each temperature increment for 2 hours. This process is repeated until self-heating is detected, the battery pack goes into thermal runaway, or the temperature reaches 200°C above the battery pack's operating temperature.

#### 8.2.3.4 *Cycling without Thermal Management*

Most EV battery packs utilize a cooling system (air or liquid cooling) to manage the cell temperature and the temperature distribution in the battery pack. A number of standards have tests that characterize the response of the battery pack in the event of a failure of the cooling/thermal management system. As an example, failure in the cooling system test in UL 2580 requires a fully charged battery pack to be conditioned at its maximum specified ambient operating temperature for 7 hours and, while at this temperature, to be discharged at the maximum specified discharge rate until it reaches its end of discharge condition with the battery pack's cooling system disabled.

#### *8.2.3.5 Immersion Test*

The aim of the immersion test is to simulate a situation where the battery pack is submerged or where battery assemblies installed in the underbody of the vehicle are partially flooded. This test is required by a number of standards, each of which requires the battery pack to be completely submerged in a saltwater solution for a period of time or until visible effects (e.g., bubbling) have stopped. Table 8.14 lists the required immersion solution and the immersion time in the different standards for this test.

### 8.2.4 Cell Internal Short-Circuit Tests

The occurrence of an internal short-circuit in a cell is of particular concern in an EV battery as it can lead to venting or a thermal runaway of a Li-ion cell [3]. Internal short-circuits in a Li-ion cell can occur due to a number of reasons, including manufacturing imperfections, dendritic growth of lithium, and the presence of contaminants in the cell. A cell internal short-circuit can create a number of scenarios in a cell (e.g., short-circuit between anode and cathode, short-circuit between anode and copper current collector, a short-circuit between aluminum current collector and copper current collector). The response of a cell to each of these short-circuit conditions is different.

**Table 8.14**
The Immersion Solution and Time
for the Immersion Tests in the Different Standards

| Standard | Immersion Solution | Immersion Time (Hours) |
|---|---|---|
| UL 2580 | 5% by weight NaCl in water | 1 |
| SAE J2464 | Nominal composition of sea water | 2 |
| SAE J2929 | 5% by weight NaCl in water | 2 |
| ISO 6469-1 | 3.5% to 5% by weight NaCl in water | 2 |
| FreedomCAR | Nominal composition of sea water | 2 |

Cell internal short-circuit tests are typically performed at a cell level and aim to understand the response of the cell to an internal short-circuit condition. Many different tests have been devised to simulate a cell internal short-circuit condition. These include:

- The forced internal short-circuit test specified in IEC 62133 involves the dismantling of a cell and the insertion of a sliver of metallic nickel between the cell's positive and negative electrodes. The cell is then reassembled and a pressing force is applied until a cell internal short is achieved.
- The penetration test specified in SAE J2464, which involves driving a metallic nail or rod through the cell at a prescribed speed.
- The dent/pinch test, which uses two spheres to apply concentrated coaxial loading on the sides of a cell with a servo-hydraulic mechanical testing machine used to control the applied pressure to the cell during the test.

## 8.3 EV CRASHWORTHINESS REQUIREMENTS

As discussed earlier in the chapter, FMVSS are prescriptive US federal regulations specifying the design, construction, performance, and durability requirements for motor vehicles. These safety standards establish minimum performance requirements for manufacturers and the equipment used to make a vehicle. FMVSS are regulated by the NHTSA, which is part of the US Department of Transportation (DOT). The FMVSS are the US counterpart to the United Nations regulations developed by the World Forum for Harmonization of Vehicle Regulations. The Code of Federal Regulations (CFR) Title 49 for Transportation governs the domestic transportation of hazardous materials for all forms of transportation to, from, and within the United States. Part 500-599 of Title 49 CFR addresses requirements from the NHTSA and the DOT. Specific requirements for electric powered vehicles are detailed in Standard number 305 of part 571. The standard:

- Specifies requirements for limitation of electrolyte spillage and retention of electric energy storage/conversion devices during and after a crash, and protection from harmful electric shock during and after a crash and during normal vehicle operation.

- Applies to passenger cars and to multipurpose passenger vehicles, trucks, and buses with a GVWR of 4,536 kg or less that use electrical propulsion components with working voltages more than 60 VDC or 30 VAC, and whose speed attainable over a distance of 1.6 km on a paved level surface is more than 40 km/h.

The standard contains a number of tests that must be performed on the vehicle with its battery at a SOC of not less than 95%. The required tests are shown in Table 8.15 [4].

Each tested vehicle must meet the following requirements to be compliant with this standard:

- Have no more than 5 liters of electrolyte leakage from the battery spill outside the passenger compartment and have no visible trace of electrolyte spill into the passenger compartment.
- Have the battery remain attached to the vehicle by at least one component anchorage, bracket, or any structure that transfers load from the battery to the vehicle structure.
- If the battery is located outside the passenger compartment, it shall not enter the passenger compartment.
- The battery must maintain electrical isolation after the test and demonstrate protection against electrical shock by direct and indirect contact.

**Table 8.15**
Part 571.304 Standard No. 305 Required Tests

| Test | Protocol |
|---|---|
| Frontal Barrier Crash | The vehicle, when traveling longitudinally at any speed up to and including 48 km/h, impacts a fixed collision barrier that is perpendicular to the line of travel of the vehicle or at any angle up to 30° in either direction from perpendicular to the line of travel of the vehicle |
| Rear Moving Barrier Impact | The vehicle is impacted in the rear by a barrier that is moving at any speed up to and including 80 km/h |
| Side Moving Deformable Barrier Impact | The vehicle is impacted from the side by a barrier that is moving at any speed up to and including 54 km/h |
| Post-Impact Static Rollover | The vehicle is rotated on its longitudinal axis to each successive increment of 90° after the frontal barrier, the rear-moving barrier impact, and the side-moving deformable barrier impact test |

*From:* [4].

## 8.4 UNIQUE REQUIREMENTS IN CHINA

The three main standards that govern EVs in China are:

- GB/T 18384-2020: Requirements for Electric Vehicle Safety;
- GB/T 38031-2020: Requirements of the Safety of Batteries Used in Electric Vehicles;
- GB/T 38032-2020: Requirements for Safety in Electric Buses.

These mandatory standards were published in May 2020 and came into effect on January 1, 2021. The requirements and tests in these standards are not very different from the tests discussed in Section 8.2. However, these standards also have a unique requirement that is not part of any of the standards in the United States or Europe. The standard requires the vehicle to provide a warning to the driver in the event of a thermal event in the vehicle's battery when the vehicle is being driven. In addition, the standard requires the vehicle to provide this warning in advance to allow egress 5 minutes prior to the presence of a hazardous situation inside the passenger compartment caused by thermal propagation triggered by an internal short-circuit leading to a single-cell thermal runaway such as fire, explosion, or smoke. This requirement is deemed to be satisfied if the thermal propagation does not lead to a hazardous situation for vehicle occupants.

Similar requirements are being proposed for EVs in Europe [5]. The proposed requirements in Europe go further and require a warning to be provided to the driver under the following conditions:

- An operational failure of one or more aspects of vehicle control(s) that manage the safe operation of the battery;
- A significant thermal event occurs internal to the RESS;
- When the battery is at a low-energy state.

## 8.5 PASSIVE PROPAGATION RESISTANCE

Designing battery packs for EV applications requires an understanding of the risks associated with thermal runaway. Thermal runaway reactions can occur with cells or batteries of any chemistry, including lead acid and nickel metal hydride batteries. Thermal runaway in a Li-ion cell occurs when the internal reaction rate in the cell increases to the point that more heat is being generated in the cell than can be withdrawn, leading to a further increase in both reaction rate and heat generation. Eventually, the amount of generated heat can be great enough to lead to the combustion of the battery as well as materials in proximity to the battery. Although rare, thermal runaway reactions in Li-ion cells do occur in the field and can be due to a number of reasons. These can include the development of cell internal short-circuits due to localized heating, highly localized mechanical damage, or latent manufacturing defects that can become active as the cell ages.

When a Li-ion cell undergoes a thermal runaway reaction, it may emit smoke, flaming liquid electrolyte and generate sufficient heat to lead to the potential combustion of material in close proximity to the cell. The failure of a single cell taken on its own may have little impact on the safety or performance of a high-energy system. However, depending on a number of factors including the cell chemistry, the cell SOC, the cell geometry, and the module or battery pack architecture, a single cell may be able to transfer sufficient heat to an adjacent cell to cause the neighbor to exceed its thermal stability limits and also undergo a thermal runaway reaction. In this way, a thermal runaway of a single cell can propagate through an entire battery pack [6]. The extent of the resulting hazard associated with a single-cell thermal runaway in an EV battery pack depends upon the likelihood that the thermal runaway reaction propagates to adjacent cells or compromises other battery or vehicle systems to create a hazard. Since this worst-case scenario can result in the release of a significant amount of energy, EV battery packs are often designed to be robust against this cascading propagation (designed with passive propagation resistance).

Propagation-resistant designs can use a combination of cell selection, spacing of cells, and interstitial materials to build passive propagation resistance (PPR) into the battery pack. As an example, NASA's manned spaceflight programs have used multicell Li-ion battery packs that utilize aluminum heat sink arrays that provide both separation of the cells and an additional path for the heat to flow to achieve PPR [7]. EV battery packs that use cylindrical cells often achieve PPR by reducing the contact area between the cells, which minimizes the heat transfer from cell to cell within a battery pack. However, often cited disadvantages of cylindrical cells as a candidate for EV battery packs include the geometric inefficiency of cylindrical cells, which limit the volumetric energy density of the battery pack as significant space is lost in the pack design. The use of polymeric pouch cells overcomes this disadvantage and allows for a lower mass and volume compared to the metal cell cans, thus offering comparatively higher-energy density at the system level. However, stacking polymeric pouch cells in close proximity to achieve this higher system level energy density makes achieving propagation resistance harder as compared to battery packs with cylindrical cells. PPR in stacked polymeric pouch-type cell battery packs can be achieved using a variety of different methods. One method involves reducing the available energy (by reducing the state of charge of the cells). Reducing the available energy in the battery pack is effective in preventing or partially reducing cascading propagation as the heat released of the failed cell is not sufficient to cause sufficient heat transfer to neighboring cells even if the cells are in thermal contact over a large surface. Studies have demonstrated that reducing the SOC to 80% or lower significantly reduces the risk of thermal runaway and thermal runaway propagation in pouch type cell battery packs [6]. Another strategy often employed to achieve passive propagation resistance involves the use of heat-absorbing aluminum or copper plates between cells. The plates provide an additional thermal mass to dissipate heat release, hence reducing the likelihood of propagation. One downside of using plates to achieve propagation resistance is the addition of a significant increase in the battery pack weight and volume if this strategy is applied to a pack with hundreds of cells. Other techniques such as the use of mica sheets are also commonly used to provide PPR.

Evaluating the effectiveness of the infrastructure used to achieve PPR in a battery requires conducting tests at the module and battery pack levels. Determining the initiation method for a single-cell thermal runaway during these tests is critical. The selected initiation method should best represent the behavior expected from a spontaneous field failure of a single battery cell, such as failure due to an internal short circuit. In addition, the selected initiation method should avoid [8]:

- Impacting the PPR infrastructure such as blocking normal exhaust gas flow or adding new and significantly different exhaust paths;

- Preheating neighbor cells beyond what would be expected from the initiator heating due to a single-cell field failure;
- Compromising the electrical isolation of the cell to any surroundings;
- Affecting the thermal boundary conditions around the initiator cell due to the addition of conductive or insulating materials that are not part of the battery pack design.

Single-cell thermal runaway during testing can be initiated with a variety of methods, and no single method is appropriate for all cell chemistries and cell form factors. It is not uncommon for multiple methods to be tried during testing to identify a method that reliably induces cell thermal runaway:

- A thermal initiation method is the most common method used to achieve a single-cell thermal runaway condition and relies on applying an externally powered heater to an individual cell. This method requires special care to limit any heating of adjacent cells during the test.
- Mechanical initiating methods sometimes attempt to simulate cell internal faults and include techniques such as nail penetration, blunt object crushing, or pinch tests. These methods result in minimum additional energy added to the system and provide a fast time to thermal runaway. However, these methods are experimentally difficult to execute at a battery pack level and can also introduce additional gas exhaust paths, alter the shape of the cell boundary, and affect the heat transfer to neighboring cells in other ways.
- Electrical initiating methods have involved cell overcharge as a way of inducing a cell thermal runaway reaction. However, this method is criticized for resulting in a reaction that is more energetic than would occur in the field that can damage surrounding cells in a nonrepresentative manner.
- Other methodologies proposed in the literature to trigger a single-cell thermal runaway include the introduction of electrode defects during the cell manufacturing process. As an example, the National Renewable Energy Laboratory (NREL) has run tests on cells that were manufactured with a wax device in-

serted between the electrode layers of the pouch cell. The wax device was heated and used to trigger an internal short-circuit during the tests. However, cells manufactured with internal defects present other challenges such as transportation, safe charging in preparation for the tests, and installation of these cells in the battery pack for the testing [9].

## 8.6 CHARGER SAFETY REQUIREMENTS

As discussed in Chapter 4, EV chargers[2] are designed to supply large currents ranging from tens of amps for Level 1 and Level 2 chargers to hundreds of amps for DC fast chargers. The large operating current and voltage introduce several potential hazards including but not limited to shock, fire, and burn hazards to the user as well as the potential for damage to the charging equipment and any connected EVs in the event of a failure. As such, EV chargers are designed to uphold minimum safety and functional requirements for safe operation in the field. EV charger standards can be generally divided into the following four categories [10–12]:

1. Electric vehicle supply equipment (EVSE) and on-board/off-board charger standards;
2. Charging connectors, inlets, and plug standards;
3. EV charging-related communication standards;
4. Safety and security standards.

Table 8.16 provides a listing of the EV charger-related IEC, ISO, and SAE standards that are implemented within North America, the EU, Japan, and South Korea. Similar to the battery-related standards, China has implemented its own unique requirements for EV chargers. These requirements are detailed in a series of GB and GB/T standards.[3] These include:

---

2. Also called electric vehicle supply equipment (EVSE). EVSEs are also commonly called EV charging stations, electric recharging points, or charging points.

3. Guobiao standards, or GB standards, are the Chinese national standards issued by the Standardization Administration of China (SAC), the Chinese National Committee of the ISO and IEC. GB stands for Guobiao, Chinese for national

**Table 8.16**
EV Charger Standards

| Standard | Application | Title | Scope |
|---|---|---|---|
| IEC 61851-1 | EVSE and on-board/off-board charger | Electric Vehicle Conductive Charging System – Part 1: General Requirements | General requirements relating to operating conditions of the EV supply equipment, specifications for connection between the EV and EV supply equipment, and electrical safety |
| IEC 61851-21-1 | | Electric Vehicle Conductive Charging System – Part 21-1 Electric Vehicle On-Board Charger EMC Requirements for Conductive Connection to AC/DC supply | Electromagnetic compatibility (EMC) requirements for on-board chargers |
| IEC 61851-21-2 | | Electric Vehicle Conductive Charging System – Part 21-2: Electric Vehicle Requirements for Conductive Connection to an AC/DC Supply – EMC Requirements for Off-Board Electric Vehicle Charging Systems | EMC requirements for off-board chargers |
| IEC 61851-23 | | Electric Vehicle Conductive Charging System – Part 23: DC Electric Vehicle Charging Station | General requirements for the control communication between an EV and a DC charger |
| SAE J1772 | | SAE Electric Vehicle and Plug in Hybrid Electric Vehicle Conductive Charge Coupler | General requirements to facilitate conductive charging in North America |
| GB/T 18487.1 | | Electric Vehicle Conductive Charging System – Part 1: General Requirements | General requirements for EV supply equipment |
| GB/T 18487.2 | | Electric Vehicle Conductive Charging System – Part 2: EMC Requirements for Off-Board Electric Vehicle Supply Equipment | EMC requirements for off-board chargers |
| GB/T 18487.3 | | Electric Vehicle Charging System AC/DC Electric Vehicle Charging Station | General requirements for conductive connection to AC/DC charging stations |

- GB/T 18487.1 – 2015: Electric Vehicle Conductive Charging System – Part 1: General Requirements;

standard. Mandatory standards in China are prefixed "GB." Recommended standards are prefixed "GB/T."

**Table 8.16** (continued)

| Standard | Application | Title | Scope |
|---|---|---|---|
| IEC 62196-1 | Charging connectors, inlets, and plugs | Plugs, Socket-Outlets, Vehicle Connectors and Vehicle Inlets – Conductive Charging of Electric Vehicles – Part 1: General Requirements | General requirements for connectors, inlets, plugs, and cable assemblies used for conductive charging systems |
| IEC 62196-2 | | Plugs, Socket-Outlets, Vehicle Connectors and Vehicle Inlets – Conductive Charging of Electric Vehicles – Part 2: Dimensional Compatibility Requirements for AC Pin and Contact-Tube Accessories | Dimensional requirements for basic interface accessories for AC vehicle supply |
| IEC 62196-3 | | Plugs, Socket-Outlets, Vehicle Connectors and Vehicle Inlets – Conductive Charging of Electric Vehicles – Part 3: Dimensional Compatibility Requirements for DC and AC/DC Pin and Contact-Tube Vehicle Couplers | Dimensional requirements for basic interface accessories for DC and AC/DC vehicle supply |
| SAE J1772 | | SAE Electric Vehicle and Plug In Hybrid Electric Vehicle Conductive Charge Coupler | Dimensional requirements for basic interface accessories for AC, DC, and AC/DC vehicle supply |
| GB/T 20234.1 | | Connection Set for Conductive Charging of Electric Vehicles – Part 1: General Requirements | General requirements relating to testing methods and testing specifications for a conductive charging connection |
| GB/T 20234.2 | | Connection Set for Conductive Charging of Electric Vehicles – Part 2: AC Charging Coupler | Specifies parameters and dimensions for AC charging coupler for conductive charging |
| GB/T 20234.3 | | Connection Set for Conductive Charging of Electric Vehicles – Part 3: DC Charging Coupler | Specifies parameters and dimensions for DC charging coupler for conductive charging |

- GB/T 18487.2 – 2017: Electric Vehicle Conductive Charging System – Part 2: EMC Requirements for Off-Board Electric Vehicle Supply Equipment;
- GB/T 18487.3 – 2001: Electric Vehicle Charging System AC/DC Electric Vehicle Charging Station;
- GB/T 20234.1 – 2015: Connection Set for Conductive Charging of Electric Vehicles – Part 1: General Requirements;

**Table 8.16** (continued)

| Standard | Application | Title | Scope |
|---|---|---|---|
| IEC 61851-24 | Communications | Electric Vehicle Conductive Charging System – Part 24: Digital Communication Between a d.c. EV Charging Station and an Electric Vehicle for Control of d.c. Charging | Communication standards between an EV and DC charging station |
| ISO 15118 series | | Road Vehicles – Vehicle to Grid Communication Interface | Communication interface for bidirectional charging and discharging of EVs |
| SAE J2847/1 | | Communication for Smart Charging of Plug-in Electric Vehicles Using Smart Energy Profile | Communications between EVs, the grid, chargers, charger networks, and customers |
| SAE J2847/2 | | Communication Between Plug-In Vehicles and Off-Board DC Chargers | |
| SAE J2847/3 | | Communication for Plug-In Vehicles as a Distributed Energy Source | |
| SAE J2847/5 | | Work in Progress – Communication Between Plug-In Vehicles and Customers | |
| SAE J2931/1 | | Digital Communications for Plug-In Electric Vehicles | Communications between plug-in EVs, the EV supply equipment, and the utility or service provider |
| SAE J2931/4 | | Broadband PLC Communication for Plug-In Electric Vehicles | |
| ISO 17409 | Safety and security | Electrically Propelled Road Vehicles – Conductive Power Transfer – Safety Requirements | Electric safety requirements for conductive connections between and EV and a charging station |
| IEC 60529 | | Degree of Protection Provided by Enclosures (IP Code) | Defines degrees of moisture and dust ingress protection as well as protection against access to potentially hazardous components |
| IEC 60364-7-722 | | Low-Voltage Electrical Installations – Part 7-722: Requirements for Special Installations or Locations – Supplies for Electric Vehicles | Relating to best practices for EV charging equipment installation |
| IEC 27000 | | Information Security Management | Best practices relating to cybersecurity and data protection |

- GB/T 20234.2 – 2015: Connection Set for Conductive Charging of Electric Vehicles – Part 2: AC Charging Coupler;

- GB/T 20234.3 – 2015: Connection Set for Conductive Charging of Electric Vehicles – Part 3: DC Charging Coupler;
- GB/T 27930 – 2015: Communication Protocols between Off-Board Conductive Charger and Battery Management System for Electric Vehicle.

### 8.6.1 EVSE and On-Board/Off-Board Charger-Related Standards

#### *8.6.1.1 IEC 61851*

IEC 61851 is the de facto standard used to evaluate EVSE and on-board chargers, especially in Europe. This standard is divided into multiple parts that include mechanical, electrical, communications, EMC, and performance requirements for both AC and DC charging stations. IEC 61851-1 (Electric Vehicle Conductive Charging System – Part 1: General Requirements) [13] delineates the general requirements for the supply of energy to EVs. This part of the standard applies to EV supply equipment for charging electric road vehicles with a rated supply and rated output voltage of up to 1,000 VAC or up to 1,500 VDC. Specific requirements for DC chargers are detailed in IEC 61851-23 (Electric Vehicle Conductive Charging System – Part 23: DC Electric Vehicle Charging Station).

Aspects specifically covered by IEC 61851-1 include:

- Characteristics and operating conditions of the EVSE;
- Specifications of the connection between the EVSE and the EV;
- Requirements for electrical safety for the EVSE.

IEC 61851-1 also includes additional requirements if the equipment is designed to operate in specific environments or conditions. These include:

- EVSE located in hazardous areas where flammable gas or vapor and/or combustible materials, fuels, or other combustible or explosive materials are present;
- EVSE designed to be installed at an altitude of more than 2,000m;
- EVSE intended to be used on board ships.

Some of the tests, evaluations, and considerations included in IEC 61851-1 and IEC 61851-23 relate to:

- Protection against electric shock through specific requirements on protection against access to energized components, disconnection of plugs when equipment is energized, and fault protection.
- Cable assembly requirements such as the electrical rating, strain relief, dielectric withstand characteristics, and construction requirements.
- Construction requirements including clearances and creepage distances, protection against dust and moisture ingress (IP degrees), insulation/isolation resistance, acceptable temperature specifications during operation under varying external conditions (damp heat functional test, minimum temperature functional test), and mechanical strength.
- Overload and short-circuit protection requirements.
- Control pilot function requirements.
- Marking and instructions requirements including the requirement for installation and user manuals, appropriate marking of the charger and charging cable assemblies, and the durability of these markings (durability test for marking).

IEC 61851-1 describes four charging modes used for AC and DC chargers as shown in Table 8.17 [13–15].

Due to the significant levels of electromagnetic interference (EMI) created by power conversion devices within charging stations (such as AC-DC and DC-DC converters), are also required to comply with EMC regulations and achieve stringent EMI thresholds. The latter aids in mitigating the propagation of interference outside of the EVSE that could potentially affect nearby communications or charging performance. IEC 61851-21-1 (Electric Vehicle Conductive Charging System – Part 21-1: Electric Vehicle On-Board Charger EMC Requirements for Conductive Connection to an AC/DC Supply) [16] and IEC 61851-21-2 (Electric Vehicle Conductive Charging System – Part 21-2: Electric Vehicle Requirements for Conductive Connection to an AC/DC Supply – EMC Requirements for Off-Board Electric Vehicle Charging Systems) [17] cover EMC requirements for on-board and

**Table 8.17**
Charging Modes Defined by IEC 61851-1

| Charging Mode | Description |
|---|---|
| Mode 1 | The EV is directly connected to an AC outlet. This is the simplest mode of charging; however, due to the lack of protective devices, it is restricted in various countries and not allowed in others. This mode of charging is usually adequate for smaller-scale battery packs such as those used in electric bikes and scooters. The maximum current for this mode is 16A, with a voltage of 250V single-phase or 480V three-phase. |
| Mode 2 | The EV is connected to an AC outlet through a cable that incorporates an in-cable control and protection device (IC-CPD). The IC-CPD performs safety functions such as ground fault, over-current, and over-temperature monitoring. Although this mode of charging does contain additional protection as compared to Mode 1, its use is still restricted in various countries including Italy, the United States, Canada, Switzerland, Denmark, France, and Norway. The maximum current for this mode is 32A, with a voltage of 250V single-phase or 480V three-phase. |
| Mode 3 | The EV is connected to a dedicated AC charging station allowing bidirectional communications between the charger and the EV. Charging stations operating in this mode contain various safety protective devices. Like Mode 2, the maximum current for this mode is 32A with a voltage of 250V single-phase or 480V three-phase. |
| Mode 4 | The EV is connected to a dedicated DC charging station allowing directional communications between the charger and the EV. As described in Chapter 4, for a DC charger, the current and voltage conversion occurs within the charging station as opposed to the on-board charger within the EV. The maximum current for this mode is 200–400A with a voltage of 400–600V. |

*From:* [13–15].

off-board charging equipment, respectively, operating in any charging mode as described in IEC 61851-1 (Table 8.17).

### *8.6.1.2 SAE J1772*

SAE J1772 (Electric Vehicle and Plug in Hybrid Electric Vehicle Conductive Charge Coupler) [18] covers the general physical, electrical, functional, and performance requirements needed to facilitate AC and DC conductive charging of EV/PHEV within North America. Some of the specific operational and functional requirements in SAE J1772 relate to:

- EMC requirements:
  - Compliance with the Federal Communications Commission (FCC) regulations.[4]

4. The FCC regulates interstate and international communications through cable, radio, television, satellite, and wire in the United States.

  - Electrostatic discharge, harmonic distortion immunity, electrical fast transient immunity, voltage variations immunity, magnetic field immunity, capacitor switching transient test, and voltage surge test in accordance with UL 2231-2 (Personnel Protection Systems for Electric Vehicle (EV) Supply Circuits: Particular Requirements for Protection Devices for Use in Charging Systems).
- Installation requirements: Compliance with the National Electrical Code (NEC) – Article 625 and Canadian Electrical Code – Part 1, Section 86.
- Safety and functionality requirements:
  - Personnel protection systems;
  - Conductor cord requirements;
  - Vehicle connector interlock and latch diagnostics;
  - Connector temperature monitoring;
  - Protection against unintended reverse power flow;
  - User initiated charge termination.
- Electrical performance requirements:
  - DC output current and voltage measurement accuracy;
  - DC output current regulation;
  - DC output current ripple;
  - DC output voltage transient;
  - DC output short circuit test;
  - DC output voltage overshoot time;
  - Voltage deviation during precharge.
- Loss of communications requirements.

Similar to IEC 61851, SAE J1772 also defines four different levels of charging with specifications, as shown in Table 8.18.

### 8.6.2 Charging Connectors, Inlets, and Plugs

EV connector designs have evolved over time and a few standardized designs are used by EVSEs today. The use of these standardized connector designs allows consumers to use any EVSE, regardless of the make and model of the vehicle. A summary of EV connectors accompanied by their relevant standards is provided in Table 8.19 [18–21]. Charging connectors and plugs are handled by users during the nor-

**Table 8.18**
EV Charging Levels Defined by SAE J1772

| Charge Method | Voltage (V) | Maximum Continuous Current (A) | Maximum Output Power (kW) |
|---|---|---|---|
| AC Level 1 | 120 VAC, 1-phase | 12A | 1.44 |
| | 120 V AC, 1-phase | 16A | 1.92 |
| AC Level 2 | 208 to 240 V AC, 1-phase | ≤80A | 19.2 |
| DC Level 1 | 50–1,000 VDC | 80 | 80 |
| DC Level 2 | 50–1,000 VDC | 400 | 400 |

mal course of operating EVSE. This makes the design and assembly of EVSE connectors and plugs especially important to ensure safety under both normal operation and in the event of a failure. Various standards have been developed to evaluate these charging couplers. These standards provide requirements for safety, environmental robustness, locking and latching mechanisms, mechanical performance, and electrical performance. Additional mandatory requirements at the federal level in the United States detailed in Part 571.305 of Standard No. 305 of Title 49 CFR delineate the conditions that the connector must meet if it can be separated from its mating component without the use of a tool. The requirement includes designing the EVSE connector to meet the following conditions:

- Necessary ingress protection rating when separated from its mating component;
- A drop-in voltage to less than or equal to 60 VDC or 30 VAC within 1 second after the connector is separated from its mating component;
- The need for two distinct actions to separate from the mating component.

#### *8.6.2.1 IEC Standards*

Requirements for charging connectors, inlets and plugs in Europe are addressed by IEC 62196. Similar to IEC 61851, IEC 62196 is also divided into multiple parts that are applicable to EV plugs, socket-outlets, connectors, inlets, and cable assemblies intended for use in conductive charging systems. These include:

**Table 8.19**
EV Charging Connectors

| Standard | Connector Type | Applicable Region | Pins | Connector Schematic |
|---|---|---|---|---|
| SAE J1772/IEC 62196-2 | SAE J1772/ Type 1 | United States, Canada, Japan | L1: AC Line 1<br>L2/N: AC Line 2/ Neutral<br>PP: Proximity Pilot<br>PE: Protective Earth (Ground)<br>CP: Control Pilot | 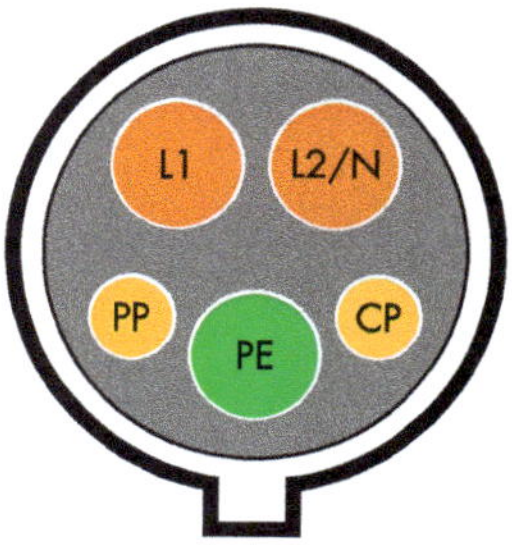 |
| IEC 62196-2 | Type 2 | Europe | L1: AC Line 1<br>L2: AC Line 2<br>L3: AC Line 3<br>N: Neutral<br>PP: Proximity Pilot<br>PE: Protective Earth<br>CP: Control Pilot | 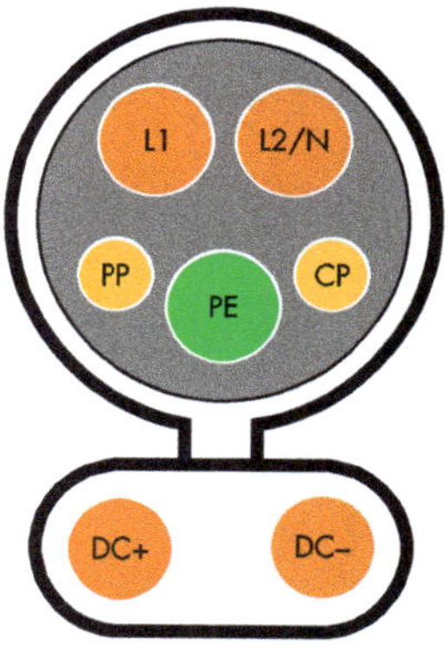 |
| IEC 62196-2 | Type 3 | Italy and France; however, it was largely replaced by the IEC 62196-2 Type 2 connector in 2013 | L1: AC Line 1<br>L2: AC Line 2<br>L3: AC Line 3<br>N: Neutral<br>PP: Proximity Pilot<br>PE: Protective Earth<br>CP: Control Pilot | 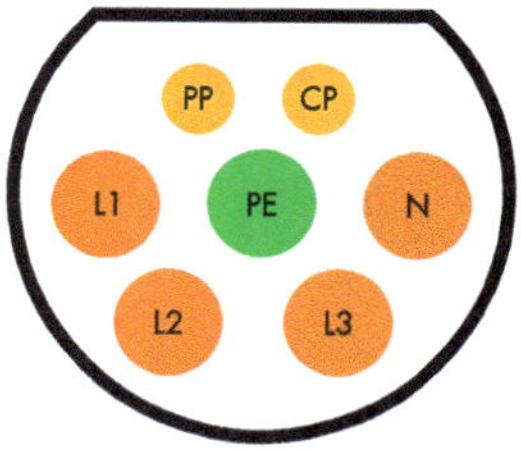 |

- IEC 62196-1 (Plugs, Socket-Outlets, Vehicle Connectors and Vehicle Inlets – Conductive Charging of Electric Vehicles – Part 1: General Requirements),[5] which provides general requirements

5. The application of a current of 80A is defined in SAE J1772 (AC Level 2) and only for applications in the United States.

**Table 8.19** (continued)

| Standard | Connector Type | Applicable Region | Pins | Connector Schematic |
|---|---|---|---|---|
| IEC 62196-2/GB/T 20234-2 | Type 2 | China | L1: AC Line 1<br>L2: AC Line 2<br>L3: AC Line 3<br>N: Neutral<br>PP: Proximity Pilot<br>PE: Protective Earth<br>CC: Charging Confirmation | 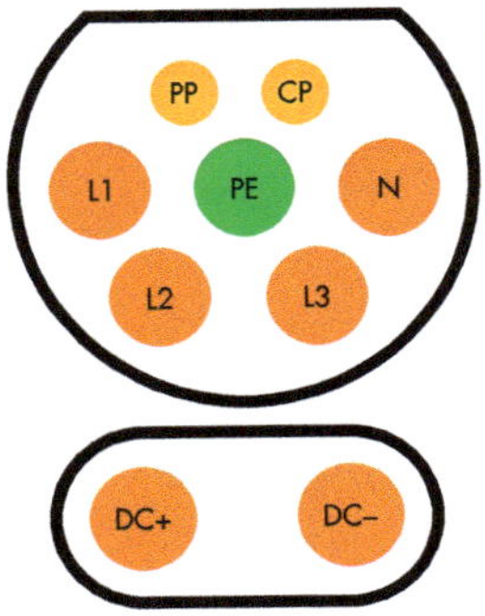 |
| SAE J1772/IEC 62196-3 | CCS Type 1 (EE) | United States, Canada, South Korea | L1: AC Line 1<br>L2/N: AC Line 2/ Neutral<br>PP: Proximity Pilot<br>PE: Protective Earth<br>CP: Control Pilot<br>DC+/DC–: DC Power | 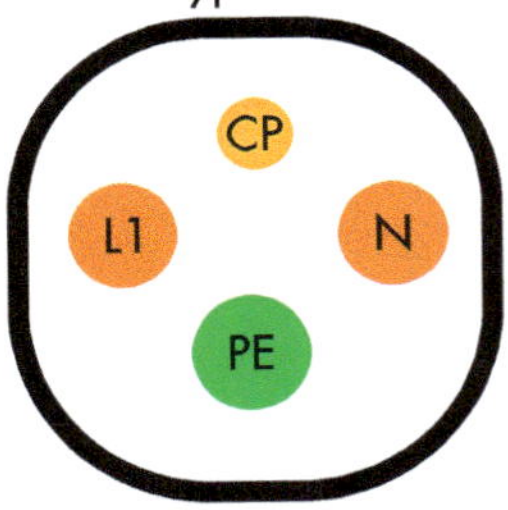 |
| IEC 62196-3 | CCS Type 2 (FF) | Europe | L1: AC Line 1<br>L2: AC Line 2<br>L3: AC Line 3<br>N: Neutral<br>PP: Proximity Pilot<br>PE: Protective Earth<br>CP: Control Pilot<br>DC+/DC–: DC Power | 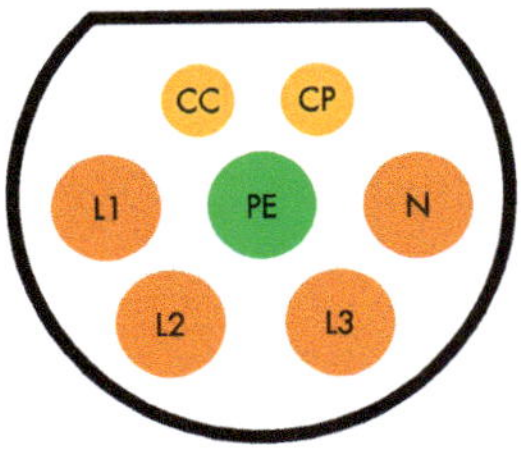 |

for the charging connector components with a rated operating voltage not exceeding:

- 690 VAC 50 Hz to 60 Hz, at a rated current not exceeding 250A;
- 1,500 VDC at a rated current not exceeding 800A.

**Table 8.19** (continued)

| Standard | Connector Type | Applicable Region | Pins | Connector Schematic |
|---|---|---|---|---|
| IEC 62196-3 | CHAdeMO (AA) | Japan | FG: Ground<br>SS1/SS2: Charge sequence signals<br>N/C: Not connected<br>DCP: Charging enable<br>DC+/DC−: DC power<br>PP: Proximity Pilot<br>C-H: CAN High<br>C-L: CAN Low | 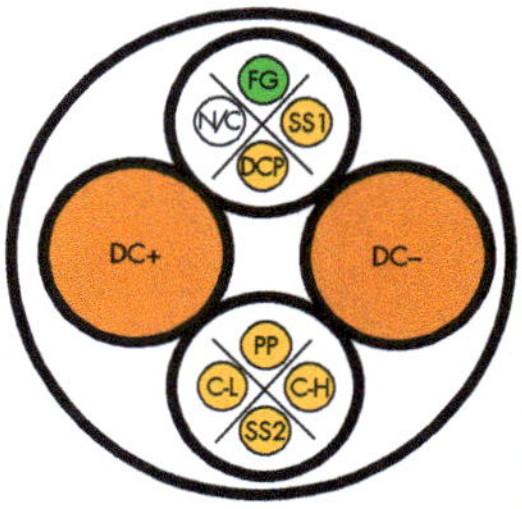  |
| IEC 62196-3/GB/T 20234-3 | BB | China | S+/S−: CAN Lines<br>CC1/CC2: Charging confirmation<br>DC+/DC−: DC power<br>PE: Protective Earth<br>A+/A−: Auxiliary DC power | 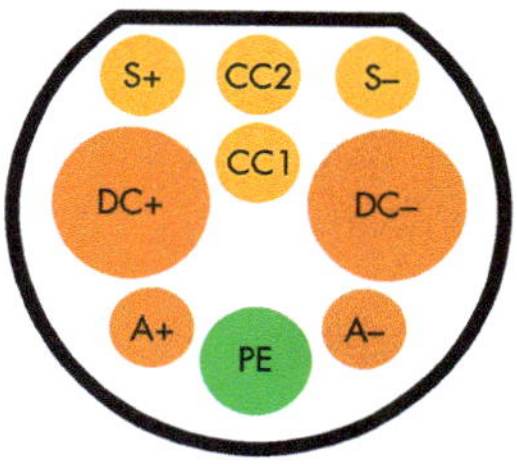  |
| North American Charging Standard (NACS)/ Tesla charging standards | Tesla | United States | DC+/L1: Positive DC power (if DC charging), AC Line 1 (single phase AC), Line 1 (split phase AC)<br>DC−/L2: Negative DC power (if DC charging), Neutral (single phase AC), Line 2 (split phase AC)<br>G: Ground<br>CP: Control Pilot<br>PP: Proximity Pilot | 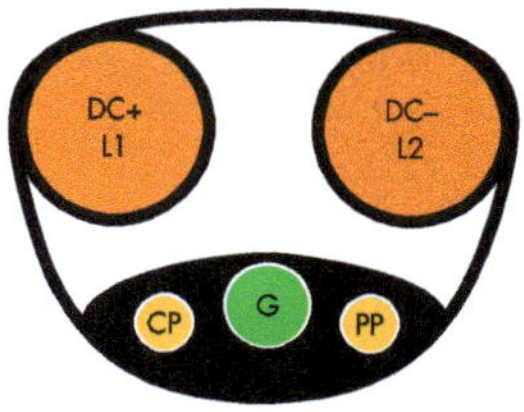  |

*From:* [18–21].

- IEC 62196-2 (Plugs, Socket-Outlets, Vehicle Connectors and Vehicle Inlets – Conductive Charging of Electric Vehicles – Part 2: Dimensional Compatibility Requirements for AC Pin and Contact-Tube Accessories) [20, 22] extends IEC 62196-1 and describes specific design requirements for AC charging vehicle inlets and vehicle connectors used in charging modes 1, 2, and 3 as defined by IEC 61851-1. Specifically, IEC 62196-2 describes 3 types of connector designs (Tables 8.19 and 8.20) that support an operating voltage not exceeding 480 VAC, 50 Hz to 60 Hz, and a rated current of up to 70A for single-phase and up to 63A three-phase.

- IEC 62196-3 (Plugs, Socket-Outlets, Vehicle Connectors and Vehicle Inlets – Conductive Charging of Electric Vehicles – Part 3: Dimensional Compatibility Requirements for DC and AC/DC Pin and Contact-Tube Vehicle Couplers) [21] applies to high-power DC interfaces and combined AC/DC interfaces used for charging of EVs in charging mode 4 as defined by IEC 61851-1. Specifically, IEC 62196-3 describes four vehicle inlet and vehicle connector configurations (Tables 8.19 and 8.21) [21].

Requirements in the various IEC 62196 parts can be divided into the following categories:

**Table 8.20**
IEC 62196-2 Charger Connector Types

| Connector Type | Supply Input | Maximum Output Current |
|---|---|---|
| Type 1 | 120V or 208/240V (single-phase) | 32A (120V) or 80A* (208/240V) |
| Type 2 | 230V (single-phase) or 400V (three-phase) | 32A (230V) or 63A (400V) |
| Type 3 | Single-phase to three-phase, 480V (max) | 63A |

*The application of a current of 80A is defined in SAE J1772 (AC Level 2) and only for applications in the United States.

**Table 8.21**
IEC 62196-3 Configurations

| Configuration | Market Name/ Designation or Relevant Standard |
|---|---|
| AA | CHAdeMO |
| BB | GB/T 20234-3 |
| EE | CCS Type 1 Connector |
| FF | CCS Type 2 Connector |

- Safety/electrical:
  - Protection against electric shock;
  - Ground connection requirements;
  - Insulation resistance and dielectric strength;
  - Breaking capacity;
  - Temperature rise;
  - Conditional short-circuit current.
- Environmental:
  - Resistance to the aging of rubber and thermoplastic material;
  - Resistance to heat and to fire;
  - Corrosion and resistance to rusting;
  - Thermal cycling;
  - Humidity exposure.
- Mechanical:
  - Dimensions;
  - Mechanical strength;
  - Contact endurance.
- Others:
  - Flexible cables;
  - EMC.

#### *8.6.2.2 SAE Standards*

SAE J1772 also contains operational, functional, and dimensional requirements for vehicle inlet and mating connectors. The J1772 connector, also known as the J plug, is designed to be used with AC charging equipment with an input voltage of 120V or 240V (AC Level 1 and AC Level 2). The 43.8-mm diameter connector is composed of the following five pins (Table 8.19):

- L1: AC Line 1.
- L2/N: AC Line 2 if used with 208 to 240 VAC; otherwise, neutral.
- PP (Proximity Pilot) or PD (Proximity Detection) is a communication line that prevents movement when the vehicle is connected to the charger. Additionally, it is also used to signal to the EV when the latch release button is pressed (i.e., when the charger is disconnected from the EV).

- PE: Protective Earth (Ground).
- CP: Control Pilot is another communication line that signals the charging level between the EV and the charger. The control pilot also activates the charge controller to begin charging when all necessary conditions are met.

Requirements in SAE J1772 can be divided into the following categories:

- Safety/electrical:
    - Maximum external surface temperature shall be as specified in UL 2251.[6]
    - Design considerations to mitigate fire, electric shock, or injury.
- Environmental:
    - Designed to withstand ambient temperatures in the range of –30°C to 50°C during operation and –40°C to 80°C during shipping or storage.
    - Specifications on the temperature rise and insulation resistance as defined in UL 2251.
    - Robustness against automotive lubricants, solvents, and fuels as specified in SAE USCAR 2.
- Mechanical:
    - The amount of force (≤75N) needed for connection and disconnection;
    - Latching to prevent accidental decoupling;
    - Locking to reduce the likelihood of unauthorized removal;
    - Design life of a minimum of 10,000 cycles of mechanical operation.
- Other:
    - Ergonomic requirements;
    - Straightforward mating of the connector and inlet free of ambiguity;
    - Tactile and/or audible feedback;

6. UL 2251 is a standard for safety for plugs, receptacles, and couplers for EVs. The standard covers plugs, receptacles, vehicle inlets, and connectors rated up to 800A and up to 600V AC or DC. The standard is intended for conductive connection systems for use with electric vehicles.

- Contact sequencing: Equipment/chassis ground contact is first make/last break and the control pilot is last make/first break.

#### *8.6.2.3 Tesla Connectors*

As shown in Table 8.19, Tesla has developed its own charging standard known as the North American Charging Standard (NACS), which applies to Tesla vehicle charging connectors and inlets. The NACS standard can support both AC and DC charging and in conjunction with the Tesla Supercharger which can deliver up to 250 kW of power [23].

### 8.6.3 EVSE Communications

One key parameter predicted to further boost EV adoption is the concept of interoperability, which refers to the compatibility of key components of the EV charging infrastructure (i.e., vehicles, charging stations, charging networks, and the electric grid) without restriction [24]. Between the EV and the charging station, interoperability means that any charger could be used with any vehicle. For the latter to occur, the EV inlet and the charger connector (hardware) must be able to connect and data (software) must be able to efficiently flow between the EV and charging station. The previous section discussing charging connectors, plugs, and inlets described the various standards that provide guidelines on the specific dimensional, electrical, environmental, functional, and mechanical requirements for the hardware. This allows for regulation and widespread use of these components around the world. Just as the hardware is standardized through these guidelines, so is the communication-related software through various standards.

- In Europe, guidelines relating to digital communications (i.e., data flow) between the charger and the EV are detailed in IEC 61851-24 (Electric Vehicle Conductive Charging System – Part 24: Digital Communication Between a d.c. EV Charging Station and an Electric Vehicle for Control of d.c. Charging) [25][7] and include requirements for:

7. IEC 61815-1 provides similar requirements for digital communication between an EV and AC charging station as well as between an EV and the management system.

   - Digital communications for charging control;
   - Bidirectional power flow;
   - Smooth and efficient data transfer through requirements for the communication circuit and communication protocol, terminating resistor, noise filer, control area network (CAN) bus, CAN transceiver, overvoltage protection for the CAN communication circuit, and CAN reception error.
- ISO 15118 defines a communication protocol between EVs and charging stations, which includes the ability for bidirectional charging/discharging of EVs (vehicle to grid (V2G)) and "Plug and Charge." The V2G capability is a two-way energy transfer between the EV and the grid to stabilize and balance the grid. The concept consists of the EV being allowed to provide energy to the grid in times of high demand and that energy being returned to the EV during low demand. The charge and discharge process of the EV would happen on a preset schedule to ensure that the EV is charged when the user needs it. The second capability encompassed by ISO 15118 is known as "Plug and Charge" where an EV would automatically start charging once connected to the charging station without the need for the user to input any additional commands or information such as payment information or charge initiation.
- As discussed in Section 8.6.2.2, SAE J1772 has specific requirements on communications between the EVSE and the EV.

On a larger scale, for interoperability within the EV charging infrastructure to continue to grow, communications between the EV and charging networks, the EV and grid, and the EV and the customers must also be considered. The current standards addressing the latter include SAE J2847/1 to SAE J2847/5 and SAE J2931/1 to SAE J2931/6.

### 8.6.4 Safety and Security Requirements

Standards used within the EV charging infrastructure are interrelated and depend on one another to describe all the necessary requirements. Most EVSE standards also reference other standards that are commonly used in a variety of industries and apply its contents within it. As noted in Table 8.16, some of these standards include:

- ISO 17409 (Electrically Propelled Road Vehicles – Conductive Power Transfer – Safety Requirements) [26], which is directly related to IEC 61851 and specifies the electric safety requirements for conductive connection between EVs and external electric circuits (charging stations). This standard breaks up its requirements into eight sections:
  - Requirements for the vehicle inlet, plug, and cable;
  - Protection against electric shock;
  - Protection against thermal incident;
  - Additional requirements for AC and DC power transfer;
  - Reverse power transfer;
  - Owner's manuals and markings;
  - Test procedures for defined electrical evaluations.
- IEC 60529 (Degrees of Protection Provided by Enclosure - IP Code) [27] applies to the classification of degrees of protection provided by electrical equipment enclosures with a rated voltage not exceeding 72.5 kV. This standard defines the degrees of protection as relevant to moisture and dust ingress into the electronic equipment as well as the potential for user access to hazardous components. Resilience toward ingress of contaminants is crucial for optimal and safe operation of electronic equipment. Contaminants, depending on their composition, in the worst case can create hazardous conditions such as a short-circuit, which can lead to the potential for a thermal event.
- IEC 60364-7-722 (Low-Voltage Electrical Installations – Part 7-722: Requirements for Special Installations or Locations – Supplies for Electric Vehicles) [28] describes specific protective measures that should be taken during EV charging equipment installation. The latter include appropriate circuit sizing, dedicated ground installations, protection against electric shock, and protection against transient overvoltage.
- ISO/IEC 27000 family (Information Security Management) [29] describes best practices for data protection. Through the concept of interoperability, the EV charging infrastructure is allowed to become more interconnected, but at the risk of cyber vulnerability due to the large amounts of freely flowing data. Cybersecurity standards such as ISO/IEC 27000 allow for cor-

porations and users to use all aspects of the rapidly growing EV infrastructure without concern of a breach to their financial information, intellectual property, or identity.

## 8.7 WIRELESS EV CHARGING STANDARDS

As described in Chapter 4, wireless power transfer (WPT) or wireless charging sets forth numerous advantages including ease of use, promise for reduction of charging times, reduction in emissions, and removal of wired connections that could present hazards. Unique concerns that need to be addressed as this new technology further integrates with EVs include [30]:

- *Electromagnetic field exposure:* Wireless charging is made possible through the transfer of energy due to rapidly changing electric and magnetic fields. As such, users must be aware and protected against electromagnetic field exposure hazards through proper labeling of restricted zones during charging and system feedback followed by subsequent charge termination if a user does encounter this hazard.
- *Inductive coil misalignment:* Unlike wired power transfer, wireless power transfer is highly dependent on the distance between the coils. Thus, the primary and secondary coils must be appropriately aligned and in close proximity to allow for efficient and safe power transfer.
- *Communications:* Through a wired connection, dedicated communications pins are used for data transfer and for charge intervention in the event of a fault condition. In a wire-free system, communications between the EV and the charging station are made wirelessly. For this to be effective, both the charging station and the EV must be able to understand the signals that are sent and respond accordingly to mitigate hazards. Additionally, wireless connections can be unreliable and thus extensive functional safety testing and validation must be performed to ensure robust communications under varying circumstances and locations.

Several standards have been developed and continue to be developed to provide guidelines and requirements for this growing EV infrastructure as summarized in Table 8.22 [31–35].

## 8.8 SUMMARY

EVs, their traction batteries, and the charging systems used to charge these batteries are subjected to a large number of tests during the de-

**Table 8.22**
Wireless EV Charging Standards

| Standard | Title | Scope |
|---|---|---|
| SAE J2954 | Wireless Power Transfer for Light-Duty-Plug-in/Electric Vehicles and Alignment Methodology | Defines acceptable criteria for interoperability, EMC, EMF, minimum performance, safety, and testing of light-duty plug-in EVs |
| SAE J2847/6 | Communication for Wireless Power Transfer Between Light-Duty Plug-In Electric Vehicles and Wireless EV Charging Stations | Minimum requirements for communication between an EV and an inductive battery charging system for wireless power transfer (WPT); extends the work in SAE J2954 by adding messages sufficient to indicate proper coil alignment has been achieved, initialize subsystems for wireless charging, ramp-up to full power, perform active wireless power transfer, and terminate the WPT session |
| IEC 61980-1 | Electric Vehicle Wireless Power Transfer (WPT) Systems – Part 1: General Requirements | The aspects covered in this standard include:<br>Characteristics and operating conditions<br>Electrical safety requirements<br>Communications between EV for WPT control<br>EMC requirements |
| IEC 61980-2 | Electric Vehicle Wireless Power Transfer (WPT) Systems – Part 2: Specific Requirements for MF-WPT System Communication and Activities | Addresses operation and functional characteristics of magnetic field wireless power transfer (MF-WPT) communications and positioning system |
| IEC 61980-3 | Electric Vehicle Wireless Power Transfer (WPT) Systems – Part 3: Specific Requirements for Magnetic Field Wireless Power Transfer Systems | Applies to off-board supply equipment for MF-WPT to EVs. The aspects covered in this standard include:<br>Characteristics and operating conditions<br>Electrical safety requirements<br>Communications requirements<br>Positioning requirements<br>EMC requirements |

*Source:* [31–35].

velopment process to ensure that the design and infrastructure of these components is reliable and can withstand the abuse conditions expected in the field. EV manufacturers and their suppliers typically use a combination of tests from industry standards together with tests specifically devised for the system in development to identify weaknesses in the design/construction and ensure that these weaknesses are addressed before the EV is put on the road. In addition, EVs and their subsystems must comply with country-specific requirements and regulations, which can impact the overall design of these subsystems.

## References

[1] Ruiz, V., et al., "A Review of International Abuse Testing Standards and Regulations for Lithium Ion Batteries in Electric and Hybrid Electric Vehicles," *Renewable and Sustainable Energy Reviews*, Vol. 81, 2018, pp. 1427–1452.

[2] Sandia National Laboratories, "FreedomCAR Electrical Energy Storage System Abuse Test Manual for Electric and Hybrid Electric Vehicle Applications," August 2006.

[3] Cai, W., et al., "Experimental Simulation of Internal Short-Circuit in Li-Ion and Li-Ion-Polymer Cells," *Journal of Power Sources*, Vol. 196, 2011, pp. 7779–7783.

[4] https://www.ecfr.gov/current/title-49/subtitle-B/chapter-V/part-571/subpart-B/section-571.305.

[5] https://wiki.unece.org/download/attachments/35782675/EVSTF09-13-TF9-01.docx?api=v2.

[6] Torres-Castro, L., et al., "Passive Mitigation of Cascading Propagation in Multi-Cell Lithium Ion Batteries," *Journal of the Electrochemical Society*, Vol. 167, 2020, p. 090515.

[7] Darcy, E., "Passively Thermal Runaway Propagation Resistant Battery Module That Achieves > 190 Wh/kg," *Sustainable Aircraft Symposium*, Redwood City, CA, May 6–7, 2016.

[8] SAE International, "Single Cell Thermal Runaway Initiation Test," Rev 1.1, September 23, 2014.

[9] Keyser, M., et al., "NREL/NASA Internal Short-Circuit Instigator in Lithium Ion Cells," *228th ECS Conference*, Phoenix, AZ, October 2015.

[10] https://www.arrow.com/en/research-and-events/articles/regulations-and-certification-for-electric-vehicles.

[11] https://www.vector.com/int/en/know-how/smart-charging/charging-standards/.

[12] https://www.ul.com/insights/electric-vehicle-onboard-equipment-and-charging-infrastructure-standards.

[13] IEC 61851-1:2017, "Electric Vehicle Conductive Charging System – Part 1: General Requirements," 2017.

[14] https://en.wikipedia.org/wiki/IEC_61851.

[15] https://www.allaboutcircuits.com/technical-articles/four-ev-charging-modes-iec61851-standard.

[16] IEC 61851-21-1:2017, "Electric Vehicle Conductive Charging System – Part 21-1: Electric Vehicle On-Board Charger EMC Requirements for Conductive Connection to an AC/DC Supply," 2017.

[17] IEC 61851-21-2:2018, "Electric Vehicle Conductive Charging System – Part 21-2: Electric Vehicle Requirements for Conductive Connection to an AC/DC Supply – EMC Requirements for Off-Board Electric Vehicle Charging Systems," 2018.

[18] SAE Electric Vehicle and Plug in Hybrid Electric Vehicle Conductive Charger Coupler J1772_201710, October 2017.

[19] IEC 61296-1:2022, "Plugs, Socket-Outlets, Vehicle Connectors and Vehicle Inlets – Conductive Charging of Electric Vehicles – Part 1: General Requirements," 2022.

[20] IEC 62196-2:2022, "Plugs, Socket-Outlets, Vehicle Connectors and Vehicle Inlets – Conductive Charging of Electric Vehicles – Part 2: Dimensional Compatibility Requirements for AC Pin and Contact-Tube Accessories," 2022.

[21] IEC 62196-3:2022, "Plugs, Socket-Outlets, Vehicle Connectors and Vehicle Inlets – Conductive Charging of Electric Vehicles – Part 3: Dimensional Compatibility Requirements for DC and AC/DC Pin and Contact-Tube Vehicle Couplers," 2022.

[22] Bopp, K., J. Bennett, and N. Lee, "Electric Vehicle Supply Equipment: An Overview of Technical Standards to Support Lao PDR Electric Vehicle Market Development," *National Renewable Energy Laboratory (NREL) Virtual Meeting*, September 18, 2020.

[23] https://cars.usnews.com/cars-trucks/advice/nacs-vs-ccs.

[24] ERPI, "Interoperability of Public Electric Vehicle Charging Infrastructure," 2019.

[25] IEC61815-24:2014, "Electric Vehicle Conductive Charging System – Part 24: Digital Communication Between a d.c. EV Charging Station and an Electric Vehicle for Control of d.c. Charging," 2014.

[26] ISO 17409:2020, "Electrically Propelled Road Vehicles — Conductive Power Transfer — Safety Requirements," 2020.

[27] IEC 60529:1989+AMD1:1999+AMD2:2013 CSV, "Consolidated Version Degrees of Protection Provided by Enclosures (IP Code)," 2013.

[28] IEC 60364-7-722:2018, "Low-Voltage Electrical Installations - Part 7-722: Requirements for Special Installations or Locations - Supplies for Electric Vehicles," 2018.

[29] https://www.iso.org/standard/iso-iec-27000-family.

[30] https://tec.ieee.org/newsletter/september-october-2014/the-unique-safety-concerns-of-wireless-charging.

[31] SAE J2954, "Wireless Power Transfer for Light-Duty Plug-In/Electric Vehicles and Alignment Methodology," J2954_202208, August 26, 2022.

[32] SAE J2847/6, "Communication for Wireless Power Transfer Between Light-Duty Plug-in Electric Vehicles and Wireless EV Charging Stations," J2847/6_202009, September 29, 2020.

[33] IEC 61980-1:2020 "Electric Vehicle Wireless Power Transfer (WPT) Systems - Part 1: General Requirements," 2020.

[34] IEC 61980-2:2023, "Electric Vehicle Wireless Power Transfer (WPT) Systems - Part 2: Specific Requirements for MF-WPT System Communication and Activities," 2023.

[35] IEC 61980-3:2022, "Electric Vehicle Wireless Power Transfer (WPT) Systems - Part 3: Specific Requirements for Magnetic Field Wireless Power Transfer Systems," 2022.

# 9

# VEHICLE-RELATED STANDARDS AND REGULATIONS

## 9.1 MOTOR VEHICLE SAFETY STANDARDS AND REGULATIONS

### 9.1.1 Federal Motor Vehicle Safety Standards

As discussed in Chapter 8, the Federal Motor Vehicle Safety Standards (FMVSS) are federal regulations within the United States that specify the minimum safety requirements to be met by motor vehicles and motor vehicle components. Under the legislative mandate of Title 49 of the US Code Subtitle VI, Chapter 301, Motor Vehicle Safety, the National Highway Traffic Safety Administration (NHTSA) is responsible for issuing and maintaining FMVSS, as well as other federal regulations, to which automobile manufacturers and motor vehicle equipment manufacturers must comply [1]. Motor vehicle safety is defined, in Title 49 of the US Code Subtitle VI, Chapter 301, as [2]: "the performance of a motor vehicle or motor vehicle equipment in a way that protects the public against unreasonable risk of accidents occurring because of the design, construction, or performance of a motor vehicle, and against unreasonable risk of death or injury in an accident, and includes nonoperational safety of a motor vehicle." As such, FMVSS provide detailed guidelines on three main safety-related performance metrics included in Title 49 of the Code of Federal Regulations (CFR) Part 571 [1, 3, 4]:

- *Crash avoidance:* This describes means by which the effects of a crash can be avoided, lessened, or mitigated. This category, also known as the 100-series, includes FMVSS No. 101 to FMVSS No. 141. A few examples include:
  - *FMVSS No. 106: Brake Hoses:* This specifies labeling and performance requirements of brake hose assemblies and fittings with the goal of reducing injuries in the event of a brake hose rupture.
  - *FMVSS No. 124: Accelerator Control Systems:* This establishes requirements for the throttle valve returning to the idle position once the driver removes his or her foot from the accelerator pedal or when a fault is detected in the accelerator pedal assembly.
  - *FMVSS No. 125: Warning Devices:* This establishes requirements for devices used to warn incoming traffic of a stopped or disabled vehicle on the road.
  - *FMVSS No. 126: Electronic Stability Control Systems:* This establishes performance and equipment requirements for electronic stability control (ESC) systems with the goal of preventing the driver from losing control of the vehicle, including vehicle rollover, during a crash.
- *Crashworthiness:* This describes the protection that the motor vehicle provides to the passengers. This category, also known as the 200-series, includes FMVSS No. 201 to FMVSS No. 227. A few examples include:
  - *FMVSS No. 203: Impact Protection for the Driver from the Steering Control System:* This establishes requirements for the steering control system with the goal of minimizing injuries to the driver on the neck, chest, and face following a crash.
  - *FMVSS No. 206: Door Locks and Door Retention Components:* This establishes performance requirements for door locks, hinges, and latches to prevent a passenger getting ejected from the vehicle in the event of a crash.
  - *FMVSS No. 208: Occupant Crash Protection:* This establishes performance requirements for the motor vehicle, in terms of forces and accelerations, for the protection of the vehicle passengers.

  - *FMVSS No. 216: Roof Crush Resistance:* This establishes strength requirements for the motor vehicle roof to prevent passenger injuries in the event of a crash, specifically a rollover.
- *Damage susceptibility and post-crash survivability:* This describes the vehicle and vehicle components performance following a crash scenario and how susceptible these components are to crash-related damage. This category, also known as the 300-series, includes FMVSS No. 301 to FMVSS No. 305. A few examples include:
  - *FMVSS No. 301: Fuel System Integrity:* This establishes spillage requirements for the fuel system to prevent and/or minimize fires following a crash. Additionally, it establishes requirements for the fuel tank against syphoning of fuel.
  - *FMVSS No. 302: Flammability of Interior Materials:* This establishes burn resistance requirements for materials used in the passenger compartment to mitigate or lessen vehicular fires because of a crash or due to a flammable source placed in the passenger compartment such as cigarettes or lighters.

A few other FMVSS within 49 CFR 571 include requirements for interior trunk release (FMVSS No. 401), platform lift systems and platform lift installations in motor vehicles (FMVSS No. 403 and FMVSS No. 404), and low speed vehicles (FMVSS No. 500).

Additionally, as discussed in Chapter 8, FMVSS No. 305 Electric-Powered Vehicles: Electrolyte Spillage and Electric Shock Protection specifically relates to requirements for passenger safety in the event of a crash involving an EV. The general requirements that must be met per this standard include [5]:

- Electrolyte spillage no more than 5 liters outside the passenger compartment with no visible trace of electrolyte spillage within the passenger compartment.
- The electric energy storage/conversion devices (i.e., the battery pack, inverters, converters) must remain adhered to the vehicle chassis and must not breach the passenger compartment in the event of a crash if located on the vehicle exterior.

- The isolation resistance must be greater than or equal to 500 Ω/V for an AC source, 100 Ω/V for an AC source conductively connected to a DC source, or 100 Ω/V for a DC source.
- Voltage measurements between chassis and the positive and negative battery pack system terminals must be less than or equal to 30 VAC or 60 VDC.
- The battery pack (high-voltage source) is IPXXB-rated.
- The resistance between exposed conductive parts and chassis is less than 0.1Ω. Additionally, the resistance between exposed conductive parts within 2.5m is less than 0.2Ω.
- The voltage between exposed conductive parts and chassis is less than or equal to 30 VAC or 60 VDC. Additionally, the voltage between exposed conductive parts within 2.5m is less than or equal to 30 VAC or 60 VDC.

The latter requirements are assessed following distinct crash scenarios, per the standard, without alteration of the vehicle during the following test sequences:

- *Frontal barrier crash:* The vehicle travels at 48 km/h and impacts a fixed collision barrier.
- *Rear-moving barrier impact:* The vehicle, equipped with test dummies, is impacted from the rear by a barrier traveling 80 km/h.
- *Side-moving deformable barrier impact:* The vehicle, equipped with test dummies, is impacted from the side by a barrier traveling at speeds up to and including 54 km/h.
- *Post-impact test static rollover:* Following the tests delineated above, the vehicle is rotated on its longitudinal axis to each successive increment of 90°.

### 9.1.2 European Requirements

The European equivalent of the FMVSS standards are the United Nations Economic Commission for Europe (UNECE)[1] standards, which aim to create a uniform system of regulations for vehicle design. Ve-

1. The United Nations Economic Commission for Europe is based in Geneva, Switzerland, and is the subordinate body of the UN Economic and Social

hicles that are compliant with the ECE regulations receive an e-Mark, which is a mandatory requirement for all automotive components and vehicles sold in the European Economic Area (EEA). By affixing the e-Mark to a product, automotive distributors and manufacturers can trade their products freely within the EEA. Nearly every European Union (EU) country has either a government agency or designated privately owned test houses that conduct testing to ensure that new models will meet all standards. Once formal approval is obtained, the automaker then issues a certificate of conformity for each vehicle manufactured, attesting that it conforms to the approved type. Once an EU member state approves a new vehicle, it can be marketed throughout the EU [6]. Currently, there are 160 different regulations that are applicable to vehicles. Some of these regulations that are directly relevant to EVs include:

- *ECE R10:* Electromagnetic compatibility requirements that detail the required EMC tests and E-marking process for electronic components, systems, and vehicles. This regulation covers:
  - Requirements regarding the immunity to radiated and conducted disturbances;
  - Requirements regarding the control of unwanted radiated and conducted emissions;
  - Additional requirements for vehicles and electronic subassemblies (ESAs) providing coupling systems for charging the rechargeable energy storage system (REESS) regarding the control of emissions and immunity from this connection between vehicle and power grid.
- *ECE R100 [7]:* Electric power train vehicle requirements, which detail the requirements that must be met by EVs. The standard is divided into two parts:
  - Part 1 includes safety requirements for vehicles with a maximum design speed exceeding 25 km/h, equipped with one or more traction motor(s) operated by electric power and not permanently connected to the grid, as well as their high-voltage components and systems that are galvanically connected

Council (ECOSOC). The UNECE is responsible for questions of economic cooperation in Europe and North America.

to the high-voltage bus of the electric power train. The requirements include:

- Protection against electric shock through both direct and indirect contact and isolation resistance requirements for both the vehicle and the coupling system for charging the REESS;
- REESS installation requirements;
- Functional safety requirements including a requirement to prohibit vehicle movement under its own propulsion system as long as the connector of the external power supply is physically connected to the vehicle inlet.

- Part 2 includes safety requirements for the REESS of road vehicles equipped with one or more traction motors operated by electric power and not permanently connected to the grid.[2] This part of the standard details several tests that must be performed on the REESS (i.e., battery) with the general requirements for each test being no leakage, cracks, fire, or explosion and for the insulation resistance to be maintained at the conclusion of the test. The tests include:
  - *Vibration test:* Sinusoidal waveform with logarithmic sweep between 7 Hz and 50 Hz and then back to 7 Hz traversed in 15 minutes. Repeated 12 times for a total of 3 hours in the vertical direction of the mount for the REESS.
  - *Thermal shock and cycling test:* This test covers changes in temperature extremes from –40°C to +60°C. The batteries are stored for 6 hours at –60°C, and then 6 hours at +60°C, for a total of 5 cycles.
  - *Mechanical shock:* 80 ms up to 28G depending on the class of vehicle, done in either a positive or negative direction or both. Typically done on a sled test machine.
  - *Mechanical integrity:* This test verifies the safety performance of the REESS under contact loads that may occur during a vehicle crash and involves a 100-kN crush using a 75-mm radius crush platen.
  - *Fire resistance:* This test verifies the resistance of the REESS against exposure to fire from outside the vehicle due

2. This regulation does not apply to REESS whose primary use is to provide power to start the engine or lighting or other vehicle auxiliary systems.

to a fuel spill. The test consists of 4 phases, which include open fuel fire preheating, direct exposure of the REESS to the flame, and indirect exposure to the flame followed by a cool down/reaction observation phase.

- *External short-circuit protection:* This test verifies the performance of the REESS protection to an external short-circuit condition that involves connecting a less than 5-mΩ load to the REESS for at least 1 hour.
- *Overcharge protection:* This test verifies the overcharge protection performance of the REESS by charging the REESS at a 1/3C rate until the REESS protection circuit interrupts or limits the charging.
- *Over-discharge protection:* This test verifies the over-discharge protection performance of the REESS by discharging the REESS at a 1/3C rate until the REESS protection circuit interrupts or limits the discharging current.
- *Over-temperature protection:* This test verifies the protection performance of the REESS against internal overheating during operation even under a cooling system failure condition. During this test, the REESS is continuously cycled at the manufacturer specified maximum continuous charge/discharge currents while the ambient air conditions are gradually increased until the temperatures have stabilized or a protection device operates to inhibit the charge and/or discharge of the REESS.

- *ECE R153:* Fuel system integrity and electric power train safety at rear-end collision requirements that evaluate the safety of the electric power train during a rear-end impact test. During the test a steel carriage/impactor assembly weighing 1,100 kg impacts the rear end of the vehicle at a speed of 50 km/h. The standard requires the vehicle to pass this test without a risk of an electric shock, with the REESS remaining attached to the vehicle and not entering the passenger compartment (if located outside the passenger compartment) and with a limit on how much electrolyte may leak from the REESS into the passenger compartment.

## 9.2 EVENT DATA RECORDER REQUIREMENTS

The deployment of airbags in any vehicle including in EVs is controlled by an airbag control module (ACM). The ACM monitors information from different sensors in a vehicle (e.g., crash sensors). If a determination is made that the vehicle has been in a crash, the ACM can deploy one or more airbags and activate seat belt pretensioners. The ACM is typically located either between the front driver and passenger seats or towards the front of the vehicle mounted to the vehicle body below where vehicle controls such as for the HVAC system are located. Figure 9.1 shows the typical input and output connections of the ACM.

ACMs are designed with an event data recorder (EDR). The purpose of the EDR in a vehicle is to record, in certain crash conditions or near-crash-like situations, data that will assist in understanding how a vehicle's systems performed leading up to the crash. The EDR is not meant to be a real-time recording device that captures and records information from all sensor signals.

In the United States, EDRs are regulated under 48 CFR part 563, which applies to passenger vehicles manufactured on or after September 1, 2010, and defines an EDR as [8]: "a device or function in a vehicle that records the vehicle's dynamic, time-series data during the time period just prior to a crash event (e.g., vehicle speed versus

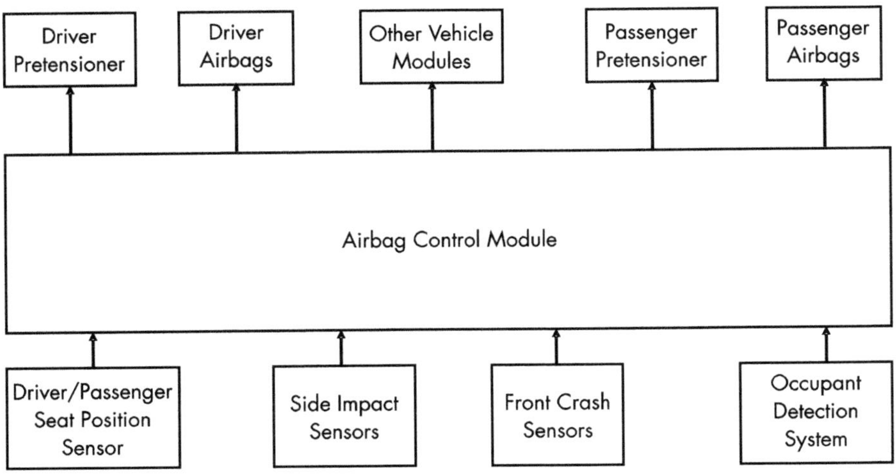


**Figure 9.1** Typical ACM system diagram.

time) or during a crash event (e.g., delta-V versus time) intended for retrieval after the crash event"[3] and specifies a "minimum core set of required data elements and accompanying range, accuracy and resolution requirements for those elements" to help to ensure that "EDRs provide the types of data most useful for the emergency medical services (EMS) community and crash reconstructionists, and in a manner that promotes consistency and comparability of these data." EDR requirements include:

- *Data element requirements:* The requirement to record a minimum set of data (15 required data elements) at a specified sample rate. Examples of these data elements include vehicle speed, driver safety belt status, and longitudinal delta-V.
- *Data accessibility requirements:* Vehicle manufacturers are required to make tools available that allow access to the data stored in EDRs. These tools can be used by law enforcement, accident investigators, and authorized technicians for retrieval and analysis.
- *Data format requirements:* The data stored in the EDR must be recorded in accordance with the range, accuracy, resolution, and filter class requirements specified in the CFR.
- *Data survivability requirements:* Vehicle manufacturers are required to ensure the survivability of the EDR data in most crashes. Compliance is checked in conjunction with crash tests required under FMVSS No. 208 (Occupant Crash Protection) and FMVSS No. 214 (Side Impact Protection).

The CFR further defines the following:

- *Time zero:* The first point in the interval where a longitudinal or lateral cumulative delta-V of over 0.8 km/h (0.5 mph) is reached within a 20-ms period.
- *Trigger threshold:* The point at which a recordable event is recognized by the EDR as suitable for further analysis. The trigger threshold is defined as "a change in velocity, in either the longitudinal or lateral direction, that equals or exceeds 8 km/h within a 150 ms interval."

3. The event data does not include audio and video data.

- *Event:* A crash or other physical occurrence that causes the trigger threshold to be met or exceeded after the end of the 500-ms period for recording data regarding the immediately previous event.
- *Pretensioner:* A device that is activated by a vehicle's crash sensing system and removes slack from a vehicle's safety belt system.

Figure 9.2 shows an example of the selected retrieved precrash EDR data that had been in an accident. As can be seen in Figure 9.2, vehicles equipped with EDRs are designed to store vehicle speed, accelerator pedal depression, service brake status (on/off), and ABS activity at a 0.5-second interval for 5 seconds leading up to the event.

## 9.3 NHTSA DEFECT INVESTIGATIONS PROCEDURE

As discussed earlier, in the United States, NHTSA is responsible for enforcing various motor vehicle safety regulations as well as ensuring compliance with these regulations. If a nonconformance or motor vehicle safety-related defect is identified, a specific procedure delineated by Title 49 of the Code of Federal Regulations (CFR) Part 574 - Standards Enforcement and Defects Investigation is followed and is headed by NHTSA's Office of Defects Investigation (ODI). The main steps that are followed include [9]:

- *Data collection:* This mainly incorporates verifying that the manufacturer certifies compliance to all safety-related standards, reviewing manufacturer's communications, collecting and screening customer complaints submitted to the Vehicle Owner's Questionnaire (VOQ) database, and, if available, obtaining information from additional sources such as online resources, news media, and local government entities.
- *Data review:* This describes reviewing the data received to identify any potential safety-related defects. Once all the data is reviewed, the risk of the identified issues is quantitatively calculated using tools such as a risk matrix to determine if an investigation should be initiated. The risk matrix is used to define the detectability of the issue, the consequences if it were to occur in the field, and the expected frequency of the issue.

**Pre-Crash Data -5 to 0 Sec (Record 1 (Table 1 of 1))**

| Time (sec) | -5.0 | -4.5 | -4.0 | -3.5 | -3.0 | -2.5 | -2.0 | -1.5 | -1.0 | -0.5 | 0.0 |
|---|---|---|---|---|---|---|---|---|---|---|---|
| Speed, Vehicle Indicated (MPH [km/h]) | 61.5 [99] | 62.1 [100] | 59.0 [95] | 49.7 [80] | 44.7 [72] | 36.0 [58] | 29.2 [47] | 21.1 [34] | 13.0 [21] | 6.2 [10] | 1.9 [3] |
| Accelerator Pedal, % Full (%) | 45 | 32 | 29 | 0 | 0 | 0 | 0 | 0 | 0 | 0 | 0 |
| Service Brake, On/Off | OFF | ON | ON | ON | ON | ON | ON | ON | ON | ON | ON |
| Engine RPM (RPM) | 1,800 | 2,700 | 2,600 | 2,300 | 1,900 | 1,500 | 1,100 | 1,100 | 1,200 | 1,200 | 1,100 |
| ABS Activity | OFF | OFF | OFF | ON | ON | ON | ON | ON | ON | ON | ON |
| Stability Control (On, Off, Engaged) | ON | ON | ON | ON | ON | ON | ON | ON | ON | ON | ON |
| Steering Input (deg) | 2.5 | -2.5 | -2.5 | -2.5 | -2.5 | -5.0 | 2.5 | -2.5 | -5.0 | 0.0 | 2.5 |

**Figure 9.2** Precrash data stored in EDR.

- *Investigation:* This phase encompasses opening a Preliminary Evaluation (PE) and formally requesting technical information from the manufacturer (Information Request (IR)). If following the initial investigation stages, it is determined that additional information, testing, or analysis is needed, the investigation is escalated to an engineering analysis (EA). An EA is largely composed of analyzing manufacturer provided data, potentially requesting technical data from peer manufacturers, conducting expert interviews, and assessing the issue through independent testing and analysis of the vehicle component. If a safety-related defect is established, following the PE or EA, and the manufacturer has not initiated a recall, NHTSA will issue a Recall Request following a review of the issue by NHTSA experts and stakeholders.
- *Recall management:* This describes assessing the manufacturers' recall plans and execution as well as monitoring recall completion.

## 9.4 ADAS REGULATIONS AND ASSESSMENT PROTOCOLS

### 9.4.1 SAE Standards

SAE International (SAE) has developed several standards for ADAS. These standards include performance-related test procedures and also address human factor-related considerations for these systems. Some of the ADAS-related SAE standards include [10]:

- *SAE J3088: Active Safety System Sensors:* This is a work in progress standard that will provide an exemplar set of acceptable requirements and tests associated with each active safety sensor technology and describe the key requirements/functionality and unique characteristics of the test targets.
- *SAE J3087: Automatic Emergency Braking (AEB) System Performance Testing:* This standard describes the equipment, facilities, methods, and procedures needed to evaluate the ability of AEB systems to detect and respond to another vehicle, in its immediate forward path, as it is approached from the rear. Three test scenarios are identified in the standard. These include:

  - A stopped lead vehicle where the vehicle being approached is stopped in the lane of the test vehicle.
  - A slower lead vehicle where the vehicle being approached is moving at a constant speed that is slower than the test vehicle.
  - A decelerating lead vehicle where the vehicle being approached is initially moving at the same speed as the test vehicle but then slows.

- *SAE J3029: Forward Collision Warning and Mitigation Vehicle Test Procedure:* This standard specifies a series of tests to evaluate the functionality of the FCW/AEB (e.g., ability to detect objects in front of the vehicle), its ability to indicate FCW/AEB engagement and disengagement, the ability of the FCW/AEB to notify the human machine interface (HMI) or vehicle control system that an object is detected under specified operating and environmental conditions, and the ability of the AEB to decelerate the vehicle to avoid impact or reduce the severity of the impact should the human operator not respond.
- *SAE J3045: Truck and Bus Lane Departure Warning Systems (LDWS) Test Procedure:* This standard specified procedures to test the functionality of the LDWS (e.g., ability to detect lane presence, and ability to detect an unintended lane departure), its ability to indicate LDWS engagement, its ability to indicate LDWS disengagement, and determine the point at which the LDWS notifies the HMI or vehicle control system that a lane departure event is detected [11].
- *SAE J3048: Driver-Vehicle Interface Considerations for Lane Keeping Assistance Systems.* This is a work-in-progress standard that will address driver-vehicle interface parameters for intervention-type LKAS.
- *SAE J2400: Human Factors in Forward Collision Warning Systems Operating Characteristics and User Interface:* This is an information report that describes elements for an FCW operator interface, as well as requirements and test methods for systems capable of warning drivers of rear-end collisions [12]. The standard does not address integration issues associated with adaptive cruise control (ACC), and, consequently, aspects of the document are considered inappropriate for an ACC system

integrated with an FCW system, making this standard less useful as most cars on the road today are equipped with both FCW and ACC systems.

- *SAE J2399: Adaptive Cruise Control Operating Characteristics and User Interface:* This standard presents the minimum requirements for safety-related elements of the operating characteristics and user interface of vehicles equipped with adaptive cruise control (ACC). It also coordinates the operating characteristics and user interface with collision warning and avoidance, along with other driver systems [13].

### 9.4.2 NHTSA Requirements

In the United States, the NHTSA requires all vehicles with Level 2 ADAS to comply with all applicable FMVSS and be certified as compliant, unless a statutory exception or exemption applies. Starting in 2021, NHTSA has required each vehicle and manufacturer and operator of Level 2 ADAS to report specific information about certain safety-related incidents involving vehicles (including prototype vehicles) operating on publicly accessible roads using Level 2 ADAS. The information must be reported within 1 to 5 calendar days after the manufacturer receives notice of the crash and must be detailed enough for NHTSA to identify crashes warranting further follow-up [14].

### 9.4.3 European Requirements

The United Nations Economic Commission for Europe (UNECE), which is part of the United Nations, has established a series of regulations for passenger safety that are relevant to ADAS. These include:

- *UN ECE R79 (Steering Equipment):* This regulation establishes provisions for advanced driver assistance steering systems that can incorporate an automatically commanded steering function (ACSF) (e.g., lane keeping or lane centering), or a correct steering function that warns the driver of any deviation from the chosen lane (LDW) and corrects the steering angle to prevent departure from the chosen lane (lane departure avoidance). Some of the test requirements in the regulation include:
  - Lane change functional test where the ACSF is activated at 10 km/h above the minimum speed at which the system is

designed to operate with another vehicle approaching from the rear.

- Minimum activation speed test where a lane change is initiated by the driver with the vehicle traveling 10 km/h below the minimum speed at which the system is designed to operate with another vehicle approaching from the rear.
- Overriding test where a lane change is initiated by the driver with the steering control firmly controlled by the driver to maintain the vehicle in the straight direction with the vehicle traveling 10 km/h above the minimum speed at which the system is designed to operate.
- Lane change procedure suppression test where a lane change procedure is initiated by the driver with another vehicle approaching from the rear with the vehicle traveling at 10 km/h above the minimum speed at which the system is designed to operate.
- Sensor performance test where the ACSF is switched on, but a lane change procedure is not initiated by the driver with the vehicle traveling at 10 km/h above the minimum speed at which the system is designed to operate and a high-volume series production motorcycle with an engine capacity not exceeding 600 cm3 approaches from the rear on an adjacent lane at 120 km/h.
- Sensor blindness test where the rear sensor is made blind with the vehicle traveling at 10 km/h above the minimum speed at which the system is designed to operate and the lane change procedure is initiated by the driver with a vehicle approaching from the rear.
- An engine start/run cycle that is divided into three phases and conducted with the vehicle traveling at 10 km/h above the minimum speed at which the system is designed to operate. The three phases include a default-off test, a check that the lane change maneuver is prevented if the system has not detected any moving object, and a lane change enabling conditions test that checks that the lane change maneuver is only possible once the system has detected a moving object at a distance equal or greater than a distance declared by the vehicle manufacturer.

- *UN ECE R 130 (LDW System):* This regulation establishes provisions for LDW systems and requires the LDWS when it is active to warn the driver if the vehicle crosses over a visible lane marking for the lane in which it is running, on a road with a directional form that varies between straight and a curve having an inner lane marking with a minimum radius of 250m, when there has been no purposeful demand to do so. The regulation provides the following test procedures:
  - Lane departure warning test where the vehicle is driven at a speed of 65 km/h into the center of the test lane and gently drifted either to the left or the right at a rate of departure between 0.1 and 0.8 m/s so that the vehicle crosses the lane marking. Tests are repeated at different rates of departure.
  - A failure detection test that simulates a LDW system failure by disconnecting the power source to any LDWS component or disconnecting any electrical connection between LDW system components.
  - A deactivation test to evaluate the warning signal generated when the LDW system is deactivated.
- *UN ECE R139 (Brake Assist Systems):* This regulation establishes uniform provisions for approving passenger cars with brake assist systems (BAS). When the driver applies maximum pressure to the brake pedal at a sudden speed and is traveling rapidly, the brake assist system works in conjunction with the antilock braking system to apply maximum pressure to the brakes. The standard defines two categories of BAS. Category A BAS detect an emergency brake condition based primarily on the brake pedal force applied by the driver, whereas Category B BAS detect emergency brake condition based primary on the brake pedal speed applied by the driver. The standard required tests to be performed to evaluate the deceleration achieved by the BAS at a test speed of 100 km/h.
- *UN ECE R152 (Advanced Emergency Braking Systems):* This regulation establishes uniform provisions for AEB systems and requires these systems to automatically detect a potential forward collision, provide the driver with an appropriate warning, and activate the vehicle braking system to decelerate the vehicle with the purpose of avoiding or mitigating the severity

of a collision in the event that the driver does not respond to the warning. The regulation provides four test procedures:

- Warning and activation test with a stationary vehicle target conducted at speeds of 20, 42, and 60 km/h;
- Warning and activation test with a moving vehicle target conducted at 30 and 60 km/h with the target traveling at 20 km/h;
- A warning and activation test with a pedestrian target conducted at 20, 30, and 60 km/h with the pedestrian target traveling in a straight line perpendicular to the subject vehicle's direction of travel at a constant speed of 5 km/h.
- Warning and activation test with a bicycle target conducted at 20, 38, and 60 km/h with the bicycle target traveling in a straight line perpendicular to the subject vehicle's direction of travel at a constant speed of 15 km/h.

## 9.5 EUROPEAN NEW CAR ASSESSMENT PROGRAM

The European New Car Assessment Program (Euro NCAP) performs a number of crash tests to assess safety in an accident. The requirements evaluate both passive safety (consequences of accidents for occupants and pedestrians) and active safety (accident avoidance measures). The vehicles are scored in the following categories [15]:

- Adult occupant protection score, which is determined from frontal impact, lateral impact, and whiplash tests and from an assessment of the measures provided for prompt and safe rescue and extrication.
  - The front impact tests include:
    - A mobile progressive deformable barrier test introduced in 2020 that simulates a crash between the test vehicle and a typical mid-size family car where the test car is driven at 50 km/h and with 50% overlap into the deformable barrier mounted on an incoming 1,400-kg trolley, also traveling at 50 km/h. Two frontal impact dummies representing the average male are seated in the front and child dummies are placed in child restraints in the rear seats.

- A full-width rigid barrier test introduced in 2015 where a vehicle is tested against a rigid barrier with full overlap at a test speed of 50 km/h with a female frontal impact dummy seated in the front driver's seat and in the rear passenger side seat.
- The lateral impact tests include:
  - A side mobile barrier test first introduced in 1997 and upgraded in 2020 where a deformable barrier mounted on a trolley is driven at 60 km/h into the side of the stationary test vehicle at right angles. A side impact dummy representing an average male is put in the driver's seat and child dummies are placed in child restraint systems in the rear.
  - A side pole test first introduced in 2001 and upgraded in 2020 where a car is propelled sideways at 32 km/h against a rigid, narrow pole. Where a vehicle is equipped with a center airbag to protect against the front seat occupants hitting each other, two average male side impact dummies are placed in the front seats. Where there is no center airbag, single dummy is placed on the driver's seat.
  - A far-side impact test first introduced in 2020 and upgraded in 2023 where a complete vehicle body is attached to a sled, which replicates the accelerations experienced by the vehicle inside and pole tests exposing a single dummy, seated on the far-side, to similar conditions as those of the full-scale tests.
  - A rear impact whiplash[4] test that was first introduced in 2009 and upgraded in 2020 that involves testing seats and head restraints on a sled using a low-speed, rear-impact dummy. Two dynamic tests are performed, representing common accident severities known to cause injuries, together with an assessment of the geometry of the front and rear head restraints.
- Rescue and extrication requirements first introduced in 2020 and upgraded in 2023 that include:

4. Whiplash injuries are associated with rapid and excessive distortion of the spine.

    - Available information for first responders in the form of rescue sheets and emergency response guides (ERG);
    - Ease of extrication such as unlocking of automatic door locking, door opening forces, seat belt unbuckling forces, compliance with ECE regulations post-crash for EV vehicles, and accuracy of eCall data;
    - Safety required such as advanced eCall, multicollision brake technology, and vehicle submergence requirements.
- Child occupant protection score based on three aspects:
    - Protection offered by child restraint systems in the frontal and side impact tests;
    - Vehicle's ability to accommodate child restraint of various sizes and designs;
    - Availability of provisions for safe transport of children in the car.
- Vulnerable road user[5] (VRU) score based on how well the vehicle protects VRU with whom they may collide. The tests evaluate the potential risks of injuries to a pedestrian's head, pelvis, and upper and lower leg and include:
    - Head impact test introduced in 1997 and upgraded in 2023 that involves a series of impact tests carried out at 40 km/h using an adult or child head impactor;
    - Pelvis impact test introduced in 1997 and upgraded in 2023 that involves a series of impact tests carried out at 40 km/h using an adult upper leg form impactor;
    - Leg impact test introduced in 1997 and upgraded in 2014 that involves a series of impact tests carried out at 40 km/h using an adult leg form impactor;
    - A series of AEB pedestrian tests[6] where for pedestrian detection three accident scenarios are tested:
        - The pedestrian is walking in the same direction as the vehicle.
        - The pedestrian is crossing a road into which the test car is turning.

5. Pedestrians and cyclists.
6. A specially designed pedestrian target is used that has articulated limbs to replicate the walking motion of a human.

    - A pedestrian is behind the reversing car.
- A series of AEB cyclist tests[7] where the following scenarios are tested:
    - A cyclist is crossing the path of the test vehicle.
    - The cyclist is ahead of the car and is approaching it from the near side.
    - The cyclist is ahead of the car and is approaching it from the near side but where the cyclist is initially obscured by vehicles parked on the near side.
    - The cyclist is approaching from the far side of the test vehicle, towards its path.
- Cyclist dooring[8] prevention test where the vehicle response to a cyclist target moving at 15 km/h on a path that is 1m from the side of a stationary car is evaluated.
- AEB/lane support motorcyclist tests that evaluate the vehicle response to seven scenarios that represent the most common crashes in Europe involving a passenger car.[9] These scenarios include:
    - Car-to-motorcyclist rear stationary;
    - Car-to-motorcyclist rear braking;
    - Car-to-motorcyclist turn across path;
    - Car-to-motorcyclist oncoming;
    - Car-to-motorcyclist overtaking;
    - Blind spot;
    - Car-to-motorcyclist crossing straight crossing path (to be introduced in 2026).
- Safety assist score that is determined from tests on driver assist technologies and includes the following tests/evaluations:

7. A specially designed cyclist target is used for the tests. A bicycle is mounted on a moving platform and the dummy cyclist is articulated to recreate the pedaling action of a human.
8. "Dooring" is an accident type where an occupant inadvertently opens the door of a parked car into the path of a cyclist approaching from behind.
9. The motorcycle target is in the form of a motorcycle dummy resembling the appearance of a motorcycle of >500 cc that is mounted on a specially designed self-propelled platform, able to steadily travel at speeds up to 80 km/h. Furthermore, the dummy features a spinning disc aimed at replicating the wheel spin and give radar systems the necessary signature for target classification.

- AEB car-car tests that involve tests with the car approaching the rear of another vehicle under the following three scenarios in which the car is approaching the rear of another vehicle:
  - Where the target vehicle has stopped;
  - Where the target vehicle is moving but at a slower speed than the test vehicle;
  - Where the target vehicle is decelerating, both gently and harshly, at different distances in front of the test car.
- Occupant status monitoring evaluations that provide scores for seatbelt monitoring and technologies that can monitor the condition of the driver through systems such as eye-monitoring sensors and identifying driving behavior that is characteristic of an impaired driver.
- Assessment of speed assist systems that:
  - Inform the driver of the present speed limit.
  - Warn the driver when the car's speed is above the set speed threshold.
  - Actively prevent the car from maintaining or exceeding the set speed.
- Lane support system tests that involve testing both lane keeping assist (LKA) and emergency lane keeping (ELK) systems against various types of road markings, including solid lines and dashed lines, and in situations where the road edge is not marked by a line. The performance is evaluated by considering the proximity of the vehicle to the edge of a lane marking or road edge at the time of intervention.

## 9.6 SUMMARY

All vehicles on the road, whether powered by a conventional gasoline engine or a battery, must comply with regulatory requirements before they are put on the road. The requirements, while similar in different regions of the world, have differences that must be considered during the design and development of the EV, as these can impact how the subsystems are designed. The addition of driver assistance features in vehicles on the road today has further increased the necessary requirements that must be met before putting a vehicle on the road.

An understanding of these requirements is a necessity to ensure that costly design modifications are not required later in the development cycle.

## References

[1] https://www.nhtsa.gov/sites/nhtsa.gov/files/fmvss-quickrefguide-hs811439.pdf.

[2] Motor Vehicle Safety, Title 49, United States Code, Chapter 301 and related uncodified provisions, May 2013, https://www.nhtsa.gov/sites/nhtsa.gov/files/mvs01092008.pdf.

[3] https://www.nhtsa.gov/sites/nhtsa.gov/files/part_c_information_standards_reqiurements_may2013.pdf.

[4] https://www.law.cornell.edu/cfr/text/49/part-571/subpart-B.

[5] https://www.ecfr.gov/current/title-49/subtitle-B/chapter-V/part-571/subpart-B/section-571.305.

[6] https://unece.org/transport/vehicle-regulations.

[7] Regulation No. 100: Uniform Provisions Concerning the Approval of Vehicles with Regard to Specific Requirements for the Electric Power Train, December 2021.

[8] Department of Transportation, NHTSA 49 CFR Part 563, Docket No. NHTSA-2006-25666, RIN 2127-AI72, Event Data Recorders, August 2006, p. 7.

[9] https://www.nhtsa.gov/sites/nhtsa.gov/files/documents/14895_odi_defectsrecallspubdoc_110520-v6a-tag.pdf.

[10] https://www.itu.int/en/ITU-T/extcoop/cits/Documents/Meeting-20210326-e-meeting/18_SAE-Advanced-Vehicle-Safety-Technology-Standards_Status-Update%20.pdf.

[11] https://www.sae.org/standards/content/j3045_201507/.

[12] https://www.sae.org/standards/content/j2400/.

[13] https://www.sae.org/standards/content/j2399_202110/.

[14] NHTSA Second Amended Standing General Order 2021-01, Incident Reporting for ADS and Level 2 ADAS, January 2001.

[15] https://www.euroncap.com/en.

# GLOSSARY

**Advanced driver assistance systems (ADAS)** Electronic systems in a vehicle that use advanced technologies to assist the driver. ADAS use sensors in the vehicle such as radar and cameras to perceive the world around it and then either provide information to the driver or take automatic action based on what they perceive. Examples of ADAS include forward collision warning (FCW), blind-spot monitoring, rear backup warning, and automatic emergency braking (AEB).

**Automatic emergency braking (AEB)** Sensors and subsystems that work together to detect when a vehicle is close to experiencing a crash. If a collision is determined to be imminent, the system automatically applies the vehicle brakes if the driver has not done so or applies supplemental braking force as necessary to avoid or mitigate the severity of the crash. (NHTSA, "NHTSA Proposes Automatic Emergency Braking Requirements for New Vehicles," May 31, 2023, https://www.nhtsa.gov/press-releases/automatic-emergency-braking-proposed-rule.)

**Anode** An electrode through which current enters any conductor of the nonmetallic class. Specifically, an electrolytic anode is an electrode at which negative ions are discharged, or positive ions are formed, or at which other oxidizing reactions occur. (IEEE, The

Authoritative Dictionary of IEEE Standards Terms, Seventh Edition, IEEE Std. 100-2000, December 11 2000.)

**Automotive Safety Integrity Level (ASIL)** Defined by ISO 26262 (International Organization for Standardization, "Road Vehicles – Functional Safety," ISO 26262:2018, Geneva, Switzerland, 2018) as "one of four levels to specify the item's (a system or an array of systems that implement a function at the vehicle level) or element's (system or part of a system) necessary requirements of ISO 26262 and safety measures (activity or technical solution to avoid or control systematic failures and to detect random hardware failures or control random hardware failures, or mitigate their harmful effects) to apply for avoiding an unreasonable residual risk (risk remaining after the deployment of safety measures), with D representing the most stringent and A the least stringent level."

**Back electromotive force (EMF)** The induced voltage and current within a motor's rotor windings created by a magnetic field that opposes the direction of the motor's stator magnetic field.

**Battery/battery pack** Composed of modules and various control/protection systems including a battery management system (BMS), cooling systems.

**Brushed DC (BDC) motor** Composed of 4 basic components; the stator, the rotor (or armature), brushes, and commutator.

**Brushless DC (BLDC) motor** Does not use brushes for commutation. Instead, the mechanical commutation in a BDC motor is replaced in a BLDC motor with switching achieved electronically using rotor position feedback sensors to determine when to switch the current.

**Battery management system (BMS)** A protection circuit incorporated in a battery pack that performs several functions including preventing the cells from operating outside their rated specifications for current, voltage, and temperature. Additionally, it also provides state of charge and state of health determination, cell balancing, and active communication between modules.

**Boost converter** A DC-DC converter that converts a lower input voltage to a higher output voltage. Also known as a step-up converter.

**Buck converter** A DC-DC converter that converts a higher input voltage to a lower output voltage. Also known as a step-down converter.

**Calendar life** The expected lifetime of a battery/cell when sitting, typically in storage or unused, at a given temperature and state of charge. (Santhanagopalan, S., et al., *Design and Analysis of Large Lithium-Ion Battery Systems,* Norwood, MA: Artech House, 2015.)

**Cathode** An electrode through which current leaves any conductor of the nonmetallic class. Specifically, an electrolytic cathode is an electrode at which positive ions are discharged, or negative ions are formed, or at which other reducing reactions occur. (IEEE, The Authoritative Dictionary of IEEE Standards Terms, Seventh Edition, IEEE Std. 100-2000, December 11, 2000.)

**Constant current-constant voltage (CCCV)** The charge profile typically used to charge Li-ion batteries. In this profile, the charger first operates in the constant current mode where the charge current is maintained at a particular constant value until the battery voltage reaches a certain voltage. The charger then transitions to the constant voltage mode where the charger output voltage is maintained at a constant value while the current decreases. Charging is terminated when the current reaches a cutoff threshold typically defined as <0.1C.

**Cell** Basic unit of an EV battery that stores electrical energy and can be charged and discharged by undergoing an electrochemical process.

**The Code of Federal Regulations (CFR)** According to Code of Federal Regulations List of Subjects, National Archives and Records Administration, April 1997, https://www.archives.gov/federal-register/cfr/subjects.html, "an annual codification of the general and permanent rules published in the Federal Register by the executive departments and agencies of the Federal Government."

**Cycle life** Related to the aging of a battery while in use during repeated charge/discharge cycles. The cycle life is dependent upon the current, voltage, and temperature values used during charging and discharging of the battery pack. The battery condition during the rest period, the time span between charge and discharge cycles, also affects the cycle life. (Santhanagopalan, S., et al., *Design and Analysis* *of Large Lithium-Ion Battery Systems,* Norwood, MA: Artech House, 2015.)

**Dynamic driving task (DDT)** According to SAE International, "Taxonomy & Definitions for Terms Related to Driving Automation Systems for On-Road Motor Vehicles," SAE J3016 2021-04, Warrendale, PA, 2021: "All of the real-time operational and tactical functions required to operate a vehicle in on-road traffic, excluding the strategic functions such as trip scheduling and selection of destinations and waypoints, and including, without limitation, lateral vehicle motion control via steering (operational), longitudinal vehicle motion control via acceleration and deceleration (operational), monitoring the driving environment via object and event detection, recognition, classification, and response preparation (operational and tactical), object and event response execution (operational and tactical), maneuver planning (tactical), enhancing conspicuity via lighting, sounding the horn, signaling, gesturing, etc. (tactical)."

**Event data recorder (EDR)** An electronic device installed in a vehicle used to record vehicle and occupant information for a few seconds before, during, and after a crash. The signals recorded by an EDR usually include precrash vehicle dynamics and system status, driver inputs, crash pulse, restraint deployment, and postcrash data such as the activation of an automatic collision notification system. (NHTSA, Event Data Recorder, February 2005, https://www.nhtsa.gov/research-data/event-data-recorder.)

**Equivalent circuit model (ECM)** A model that simulates a battery by using simple electrical components as an analog to the physiochemical principles.

**Electric vehicle supply equipment (EVSE)** Commonly called charging stations and used to supply power to an EV's battery during the charging process. EVSE systems interface with the vehicle through dedicated communications protocols to ensure that energy is delivered to the battery safely and efficiently. (NEMA, "Electric Vehicle Supply Equipment/System," https://www.nema.org/membership/products/view/electric-vehicle-supply-equipment-system.)

**Fail-safe** A system or component that automatically places itself or the larger system in a safe operating mode in the event of a failure. (IEEE, The Authoritative Dictionary of IEEE Standards Terms, Seventh Edition, IEEE Std. 100-2000, December 11, 2000.)

**Forward collision warning (FCW)** A system intended to passively warn the driver via audible, visual, and/or haptic alerts to assist in avoiding or mitigating a rear-end collision. (Forkenbrock, G., and B. O'Harra, A Forward Collision Warning (FCW) Performance Evaluation, December 2008, https://www-esv.nhtsa.dot.gov/Proceedings/21/09-0561.pdf.)

**Failure modes and effects analysis (FMEA)** A methodology designed to identify potential failure modes for a product or process, to assess the risk associated with those failure modes, to rank the issues in terms of importance, and to identify and carry out corrective actions to address the most serious concerns. (Mendias, M., S. Lele, and A. Arora, "Functional Safety & Safety Critical Systems - An Overview," SAE Technical Paper 2021-01-0157, 2021.)

**Federal Motor Vehicle Safety Standards (FMVSS)** Guidelines and regulations issued by the NHTSA to implement laws from Congress.

**Formation** The initial charging and conditioning of a Li-ion cell to activate its electrochemical properties and optimize its performance.

**Fault tree analysis (FTA)** A methodology that provides a means to logically and graphically display the paths to failure for a system or component. (Mendias, M., S. Lele, and A. Arora, "Functional Safety & Safety Critical Systems - An Overview," SAE Technical Paper 2021-01-0157, 2021.)

**Functional safety** Part of the overall safety that depends on a system or equipment operating as designed in response to its inputs. At a high level, the term functional safety refers to (Mendias, M., S. Lele, and A. Arora, "Functional Safety & Safety Critical Systems - An Overview," SAE Technical Paper 2021-01-0157, 2021) the: "detection of a potentially dangerous condition resulting in the activation of a protective or corrective device or mechanism to prevent hazardous events from arising or providing mitigation to reduce the consequences of the hazardous event."

**H-bridge** An electronic circuit that switches the polarity of a voltage applied to a load.

**Hazard and operability study (HAZOP)** A structured and systematic technique for system examination and risk management. In particular, HAZOP is often used as a technique for identifying potential hazards in a system and identifying operability problems likely to lead to nonconforming products. (Mendias, M., S. Lele, and A. Arora, "Functional Safety & Safety Critical Systems - An Overview," SAE Technical Paper 2021-01-0157, 2021.)

**Intercalation** The process in Li-ion cells where the lithium ions are reversibly inserted and removed from the electrodes during operation while no significant change occurs to the structure of the host (i.e., the electrodes during operation).

**ISO 26262** An international standard for functional safety in the automotive industry that defines guidelines for electrical and electronic systems in road vehicles.

**Jelly roll** Cylindrical lithium-ion batteries are manufactured by rolling the different battery layers into a cylindrical roll, which is then placed in a metal can. The resulting rolled spiral structure is commonly referred to as a "jelly roll."

**Lane centering assist (LCA)** Active steering of a vehicle toward the center of the lane. LCA can be activated while driving and can be overridden or deactivated by the driver at any time during the driving task. The driver remains responsible for the driving task. ("Lane Centering Assist," Bosch Mobility, https://www.bosch-mobility.com/en/solutions/assistance-systems/lane-centering-assist/.)

**Lane departure warning (LDW)** This warns the driver if straying from the marked lane lines. When the function detects that the vehicle is about to unintentionally move out of the lane, it warns the driver by means of a visual, audible, and/or haptic signal, such as steering wheel vibration. These warnings signal the driver that the vehicle is drifting off-course, allowing him or her to counter-steer accordingly. The function does not issue a warning when the driver activates the turn signal to change lanes or turn intentionally.

**Lane keeping assist (LKA)** Actively helps drivers to keep their vehicle within the lane. If the vehicle's distance to the lane markings falls below a defined minimum, the system steps in and using the vehicle's electrical power steering system counter steers to keep the vehicle in the lane. In vehicles without electric power steering, it achieves the same effect by utilizing the electronic stability program (ESP) to brake individual wheels.

**Lorentz force** In the context of motors, it is the force experienced by the current carrying windings within the rotor when in the presence of a magnetic field.

**Module** A battery assembly put into a frame by combining a certain number of cells in series and parallel combinations.

**The National Highway Traffic Safety Administration (NHTSA)** According to Federal Register, https://www.federalregister.gov/agencies/national-highway-traffic-safety-administration, this "was established by the Highway Safety Act of 1970 to help reduce the number of deaths, injuries, and economic losses resulting from motor vehicle crashes on US highways. The Administration carries out programs relating to the safety performance of motor vehicles and related equipment; administers the State and community highway safety program with the FHWA; regulates the Corporate Average Fuel Economy program; investigates and prosecutes odometer fraud; carries out the National Driver Register Program to facilitate the exchange of State records on problem drivers; conducts studies and operates programs aimed at reducing economic losses in motor vehicle crashes and repairs; performs studies, conducts demonstration projects, and promotes programs to reduce impaired driving, increase seat belt use, and reduce risky driver behaviors; and issues theft prevention standards for passenger and non-passenger motor vehicles."

**Nominal voltage** According to Recommendations on the Transport of Dangerous Goods, Manual of Tests & Criteria, Fifth Revised Edition, Amendment 1, United Nations Economic Commission for Europe, December 2011: "The approximate value of the voltage used to designate or identify a cell/battery."

**Object and Event Detection and Response (OEDR)** The subtasks of the dynamic driving task that include monitoring the driving environment (detecting, recognizing, and classifying objects and events and preparing to respond as needed) and executing an appropriate response to such objects and events.

**Passive propagation resistance (PPR)** The design of a battery pack/module that prevents the spread of thermal runaway event. PPR is designed to detect and isolate individual battery cells that have experienced a thermal incident. The design of the battery pack is focused on PPR, so if one cell fails, it will not spread to adjacent cells.

**Process hazard analysis (PHA)** A thorough, orderly, systematic approach for identifying, evaluating, and controlling the hazards of processes. The methodology establishes priority based on the extent of process hazards, the number of potentially affected individuals, the age of process and the operating history of the process. (Mendias, M., S. Lele, and A. Arora, "Functional Safety & Safety Critical Systems - An Overview," SAE Technical Paper 2021-01-0157, 2021.)

**Pulse width modulation (PWM)** A modulation technique that generates variable-width pulses to represent the amplitude of an analog signal.

**Regenerative braking** The use of electric motors to slow down and stop a vehicle. With regenerative braking, the braking system captures the kinetic energy from the braking event and transfers it into the vehicle's battery.

**Safety Integrity Level (SIL)** According to International Electrotechnical Commission, "Functional Safety of Electrical/Electronic/Programmable Electronic Safety-Related Systems - Part 1: General Requirements," IEC 61508-1, Geneva, Switzerland, 2010, "The probability of a Safety Instrumented Function (SIF) satisfactorily performing the required safety functions under all stated conditions within a stated period of time." Succinctly, SIL refers to the probability of a safety function working correctly when it is needed.

**State of charge (SOC)** A battery's SOC describes the amount of energy available in the battery at a given time. Technically, SOC is defined as the percentage of the remaining charge inside a battery to the full charge, and thus it can range from 0% to 100%.

**Switching losses** The energy dissipated as heat during the transition between the on-state and off-state of a switching component such as a MOSFET or IGBT.

**Thermal runaway** A condition where heat is generated much more rapidly that it is dissipated leading to rapidly increasing temperatures. (Babrauskas, V., Ignition Handbook: Principles and Applications to Fire Safety Engineering, Fire Investigation, Risk Management and Forensic Science, Fire Science Publishers, 2003.)

**Vulnerable road user (VRU)** A term to describe unprotected individuals such as pedestrians or cyclists, as they are at greater risk of sustaining major injuries in the event of collision with a vehicle. (National Safety Council, "Vulnerable Road Users," https://www.nsc.org/getattachment/d5babee6-582d-4e66-804f-8d06f9b021a4/t-vulnerable-road-users-147.)

# ABOUT THE AUTHORS

**Ashish Arora** received his bachelor's degree in electronics and communications engineering from the University of Canberra, Australia, in 2001, his master's degree in electrical and computer engineering from Purdue University, Indiana, in 2003, and his MBA from Indiana University in 2015. Since 2004, he has been employed at Exponent, Inc., an engineering consulting firm that specializes in the analysis and prevention of failures and accidents. He is a principal engineer in Exponent's Electrical Engineering and Computer Science practice and is based in Exponent's Phoenix, Arizona office.

Mr. Arora specializes in electrical and electronic systems and has expertise in analog and digital electronic systems, computer systems, programmable controllers, and software systems. He has also worked extensively on a wide array of lithium-ion battery systems for a variety of applications and industries including for EVs. Mr. Arora is the author of several journal and conference papers, a number of book chapters, and a book on lithium-ion battery failures in consumer electronics.

**Rita Garrido Menacho** received her bachelor's degree in physics from the Georgia Institute of Technology in 2013 and her PhD in physics from the University of Illinois at Urbana-Champaign in 2020. Dr. Garrido Menacho is a condensed matter physicist with specialties in semiconductor/superconductor nanofabrication, surface characterization, and nanoscale electrical testing.

Since 2021, Dr. Garrido Menacho has been employed by Exponent and works as an engineering and scientific consultant with a focus on failure analysis of consumer electronics, automotive electronic systems, and battery systems. Dr. Garrido Menacho's expertise includes lithium-ion battery pack quality and design evaluations as well as product design safety through electrical, thermal, and mechanical testing. Dr. Garrido Menacho is also experienced in investigations involving automotive electronic system failures and recall-related matters. Dr. Garrido Menacho has authored book chapters in the fields of energy storage system safety and compliance as well as battery management systems.

# INDEX

## Artech House Power Engineering Library

Naoki Matsumura, Series Editor

*Advanced Technology for Smart Buildings,* James Sinopoli

*The Advanced Smart Grid: Edge Power Driving Sustainability, Second Edition,* Andres Carvallo and John Cooper

*Applications of Energy Harvesting Technologies in Buildings,* Joseph W. Matiko and Stephen Beeby

*Battery Management Systems, Volume I: Battery Modelings,* Gregory L. Plett

*Battery Management Systems, Volume II: Equivalent-Circuit Methods,* Gregory L. Plett

*Battery Management Systems, Volume III: Physics-Based Methods,* Gregory L. Plett and M. Scott Trimboli

*Battery Management Systems for Large Lithium-Ion Battery Packs,* Davide Andrea

*Battery Power Management for Portable Devices,* Yevgen Barsukov and Jinrong Qian

*Big Data Analytics for Connected Vehicles and Smart Cities,* Bob McQueen

*Design and Analysis of Large Lithium-Ion Battery Systems,* Shriram Santhanagopalan, Kandler Smith, Jeremy Neubauer, Gi-Heon Kim, Matthew Keyser, and Ahmad Pesaran

*Designing Control Loops for Linear and Switching Power Supplies: A Tutorial Guide,* Christophe Basso

*Electric Power System Fundamentals,* Salvador Acha Daza

*Electric Systems Operations: Evolving to the Modern Grid, Second Edition,* Mani Vadari

*Energy Harvesting for Autonomous Systems,* Stephen Beeby and Neil White

*Energy IoT Architecture: From Theory to Practice,* Stuart McCafferty

*Energy Storage Technologies and Applications,* C. Michael Hoff

*EV Batteries, Chargers, and Subsystems,* Ashish Arora and Rita Garrido Menacho

*GIS for Enhanced Electric Utility Performance,* Bill Meehan

*IEC 61850 Demystified,* Herbert Falk

*IEC 61850: Digitizing the Electric Power Grid,* Alexander Apostolov

*Introduction to Power Electronics,* Paul H. Chappell

*Introduction to Power Utility Communications,* Harvey Lehpamer

*IoT Technical Challenges and Solutions,* Arpan Pal and Balamuralidhar Purushothaman

*Lithium-Ion Batteries and Applications: A Practical and Comprehensive Guide to Lithium-Ion Batteries and Arrays, from Toys to Towns, Volume 1, Batteries,* Davide Andrea

*Lithium-Ion Batteries and Applications: A Practical and Comprehensive Guide to Lithium-Ion Batteries and Arrays, from Toys to Towns, Volume 2, Applications,* Davide Andrea

*Lithium-Ion Battery Failures in Consumer Electronics,* Ashish Arora, Sneha Arun Lele, Noshirwan Medora, and Shukri Souri

*Lithium-Ion Battery Standards,* Jan Swart and Jody Leber

*Microgrid Design and Operation: Toward Smart Energy in Cities,* Federico Delfino, Renato Procopio, Mansueto Rossi, Stefano Bracco, Massimo Brignone, and Michela Robba

*Plug-in Electric Vehicle Grid Integration,* Islam Safak Bayram and Ali Tajer

*Power Grid Resiliency for Adverse Conditions,* Nicholas Abi-Samra

*Power Line Communications in Practice,* Xavier Carcelle

*Power System State Estimation,* Mukhtar Ahmad

*Practical Battery Design and Control,* Naoki Matsumura

*Renewable Energy Technologies and Resources,* Nader Anani

*Robust Battery Management System Design with MATLAB®,* Balakumar Balasingam

*Signal Processing for RF Circuit Impairment Mitigation in Wireless Communications,* Xinping Huang, Zhiwen Zhu, and Henry Leung

*The Smart Grid as An Application Development Platform,* George Koutitas and Stan McClellan

*Smart Grid Redefined: Transformation of the Electric Utility,* Mani Vadari

*Sustainable Power, Autonomous Ships, and Cleaner Energy for Shipping,* John Erik Hagen

*Synergies for Sustainable Energy,* Elvin Yüzügüllü

*A Systems Approach to Lithium-Ion Battery Management,* Phil Weicker

*Telecommunication Networks for the Smart Grid,* Alberto Sendin, Miguel A. Sanchez-Fornie, Iñigo Berganza, Javier Simon, and Iker Urrutia

*A Whole-System Approach to High-Performance Green Buildings,* David Strong and Victoria Burrows

For further information on these and other Artech House titles, including previously considered out-of-print books now available through our In-Print-Forever® (IPF®) program, contact:

Artech House
685 Canton Street
Norwood, MA 02062
Phone: 781-769-9750
Fax: 781-769-6334
e-mail: artech@artechhouse.com

Artech House
16 Sussex Street
London SW1V 4RW UK
Phone: +44 (0)20 7596-8750
Fax: +44 (0)20 7630-0166
e-mail: artech-uk@artechhouse.com

Find us on the World Wide Web at: www.artechhouse.com